꺼꾸로 읽어보는 우주에 관한 유쾌한 상상

중력의 기원과 새로운 우주

行憲 김 민 태

홍릉과학출판사

This book is an English and modified version of "중력의 기원과 새로운 우주" (홍릉과학출판사, Hongreung Science Publishing Co.) written in Korean (published 30.05.2019, ISBN 9791156006732).

The Origin of Gravity and the New Cosmos

An imaginary experiment on gravity and the universe

行憲 김 민 태*1)

Min Tae Kim, PhD

---

*1) Doctor of Engineering. Bachelor of Metallurgical Engineering, Seoul National University (1983), Master of Science and Technology of Korea (1985), Dr. of Hamburg University of Technology (1993). He has worked for KIA Motors and currently works at KEPCO Research Institute, KEPRI.

Putting this book into the world ...

History is a constant challenge and response. It is the definition of the history by the English historian Toynbee. This word would be also perfectly applied to the history of science. Classical mechanics or Newtonian mechanics, which dominated physics until the beginning of the 20th century, gave way to the challenge of quantum mechanics. Quantum mechanics has been at the very top of modern physics throughout the 20th century. Physics of microscopic world is well described in terms of quantum mechanics. But when you look up into the distant universe, things are different. Quantum mechanics does not work because of gravity. Gravity is hot potato that is not compatible with quantum mechanics. Newton's law of gravity along with Einstein's theory of general relativity is leading the physics of the universe. Numerous cosmic phenomena cannot be explained without gravity, and the theory of relativity explains them in sophisticated languages.

Is the physics of macroscopic world so different from that of microscopic world? If they are the same, and if they are inevitably the same, where can we find the answer to the problem? To find an invisible clue, it is often useful to think in a reverse way. We filled the empty vacuum with a very hard medium. Gravity is not an attractive force but a pushing force from the background vacuum. This is why the subtitle of this book is "a jolly imagination about gravity and the universe".

Recently, more and more books on science have been published, excluding textbooks and major books. It is something I have to appreciate as a person engaged in science and technology. Most of the current books on science are a kind of history books that introduce past scientific discoveries or theories. There are books describing the history of science from the viewpoint of "challenge and response" and there are books that simply carry information. They are things that summarize past events and thoughts. This book, titled "Origin of Gravity and New Cosmos," is a kind of science book, but it is not a history book. Though I looked into the records and ideas of the past, imagined many things, and tried to come up with new ideas logically. It is not a book that records the past, but a book of foresight telling the future. If things are verified and acknowledged, it will become another history book. It is a dream of the author.

This book summarizes the author's thoughts over the years. I used a lot of internet resources such as wikipedia.org to support my thoughts. It would be impossible to understand the contents of this book at once. For readers out there, I tried to explain unfamiliar concepts and phenomena as simply as possible. However, the introduction of some mathematics was inevitable to prove my ideas. Science could not develop unless it is denied and challenged. It would be the virtue of scientists to doubt and interpret from other view points, even if it is already well established and seemingly lacking of any faults. I hope that the process of practicing these virtues itself is appreciated apart from the right and wrong of the contents of this book. I would like to express my sincere gratitude to Mr. Woo Myung Chan, the head of Hongreung Science Publishing Co., Mr. Kim Ki-Yong and other publishers who helped me a lot.

Introduction

When the sky is clean and clear, a lot of stars are seen at night. We can talk a lot while watching the starlight. Not only in the present but also in the distant past, scientists have observed stars, scientifically analyzing their movements and trying to understand the principles. The related disciplines are called astronomy. The movement of stars were also used to tell the fortune of individuals and nations. People trusting in astrology are still seen. Among a lot of stars in the sky there are ones that brightly shine themselves like the Sun (we do not call the Sun a star, but it is academically a star), and there are planets such as Earth, Mars, Jupiter, etc. There are also meteors. Stars are telling us in their own language, namely light. Light is the fastest we know. (Neutrinos were once known to be faster than light, but it turned out to be erroneous.)[1] The Sun is away from us just within 8 minutes at the speed of light (exactly 8 minutes and 19 seconds), but there are many stars very far away from us which cannot be reached after travelling for more than a decade (or a few hundred million years) at the speed of light.

How can light move at the speed of light in the vacuum where there is nothing at all? A few hundred years ago, when the property of light was not yet known exactly, Newton insisted that light consisted of particles. In those days, scientists like Huygens experimentally showed that light is a kind of wave like a sound wave. If light is a wave, it can move only if there is fluid or any medium. Scientists who believed that light was a wave in the Newtonian era thought "Ether" as a medium, a fictitious material (fluid) that light can rely on. In 1887, A. A. Michelson and E. Morley experimentally tried to prove the existence of this hypothetical optical fluid. They developed a sophisticated optical device to observe the interference phenomenon expected to show up due to the speed difference of light in the presence of Ether. This is the famous "Michelson-Molly" experiment.

When Earth rotates around the Sun and feels the flow of "Ether" and this flow will be felt differently in the direction perpendicular to and parallel to the direction of Earth's rotation. When two beams of light move the same distance, but in the perpendicular directions, the speed of light of the two beams will be different from each other, so that a interference pattern had been expected to appear. However, unlike the expectation, no interference was observed. Though more sophisticated experiments had been performed since then, the results were the same. There was no such thing like "Ether" that light can move through. For this experiment denying the presence of Ether, Michelson won the Nobel Prize in Physics in 1907.

As a result, light travels through the vacuum by itself and has been recognized as a dual nature of wave and particle at the same time, since the particle characteristics of light was proved in the latter half of the 20th century (light particles, namely photons, contributed greatly to the emergence of quantum mechanics, the core of modern physics). Einstein proposed the theory of special relativity from the fact that light travels through the vacuum by itself and that the speed of light is invariant regardless of the state of the observer (the same velocity is measured). From this theory, Einstein mass-energy equivalence was derived, which tells mass (m) is convertible with energy (E) via the speed of light (c): $E = mc^2$. This formula fully explains the relationship between the mass difference before and after the reaction and the energy released during nuclear fusion or nuclear fusion, which further strengthened the conviction that the space in which the light travels is empty.

Is there really no medium that mediates the movement of light in the vacuum? Is the space composed only matter and nothing? Is there no energy or substance in the vacuum that we do not know? Recent trends are approving the hypothesis that the space is composed of dark matter and dark energy in addition to ordinary matter we know. There is definitely something in the background of the universe. The vacuum also has matter or energy. Ironically, Einstein is the first person to argue that there is energy or matter in the vacuum. (Why ironic? The theory of special relativity is derived from the assumption that there is nothing to hold light in the vacuum, although Einstein presupposed that something in the vacuum does not react with light at all ...) According to him, the universe will shrink due to gravity. If it does not, then there should be energy to prevent its shrinkage, and the concept of the cosmological constant was added to his theory of general relativity. This amended theory is a cosmological theory stating that the universe does not contract or expand. However, as Hubble's observations revealed that the universe is expanding, the cosmological constant was no longer accepted. With the recent recognition that the universe expand acceleratedly, however, the cosmological constant has received renewed attention and astronomical assumptions are emerging that there will be dark energy or dark matter in the universe that we do not know.

Today, the vacuum in the universe is accepted not to be an empty space, but there is only one question, how much energy does it have in itself. At present, two theories are in conflict. The vacuum has the lowest energy, namely zero point energy, 10^{-9} J/m^3 or conversely it has enormous energy predicted to be 10^{113} J/m^3. The difference is so large that physicists call it "the vacuum catastrophe" or "the cosmological constant problem". If these two theories are correct, we may introduce the following concept of the vacuum.

"The vacuum is made of a very dense medium (called the solid vacuum), but it contains zero or very little energy."

The above premise implies that the medium constituting the vacuum does not convert into energy via mass-energy equivalence. In addition to this premise, the second premise is that energy must be added to the solid vacuum to become matter. That is, the physical world of matter is determined by the interaction between the medium constituting the vacuum and the energy held in it. Stephen Battersby, editor of a British science magazine, wrote, "It is confirmed: Matter is merely vacuum fluctuations." October 16, 2008 in a news article of New Scientist (newscientist.com). Also, in 1985, Paul Davies noted that "knowing about vacuum is the key factor in approaching Theory of Everything."[2] Knowing about the vacuum should be the key to the understanding our physical world.

Contents

Abstract   13

I. Two mysterious things: light and gravity   15

   1.1. Light needs a medium for its travel?   16

   1.2. Distortion of time − Special Relativity   21

   1.3. Gravity, the very unknown force   28

   1.4. General Relativity with no gravity   37

II. Origin of gravity   45

   2.1. Universe filled with a very hard medium   46

   2.2. Light is trembling of the solid vacuum   55

   2.3. The starry night of Gogh   63

   2.4. Mass: energy stored in the solid vacuum   69

   2.5. Matter wave: the essence of the
movement of matter   76

   2.6. Gravity cannot exist on its own   83

   2.7. Deflection of light in the distorted
solid vacuum   96

   2.8. Procession of Mercury   100

III. New thinking on the universe 111

3.1. Where comes the energy of the universe? 112

3.2. The Sun in the solid vacuum 115

3.3. A cold star - Earth 123

3.4. Jupiter resembling the Sun 132

3.5. Supernova - Explosion of the compressed energy 132

3.6. Ultra-fast spinning neutron stars 157

3.7. Paradox of black holes 166

IV. For the new universe 181

4.1. Cosmic Microwave Background 182

4.2. Dark matter, part of the solid vacuum? 197

4.3. Dark energy - Energy contained in the solid vacuum? 207

4.4. Instead of the Big Bang theory 215

V. The universe is simple and clear in the solid vacuum 223

References 231

ABSTRACT

In modern physics, gravity is still an unknown physical phenomenon. Many attempts have been made to predict and observe the presence of gravitational waves based on quantum gravity theory, but nothing has been found yet. Of course, there are news that they have been discovered. The magnitude of this gravitational wave is so small that it compares the size of a hydrogen atom to that of the Sun. In reality, it means that the measured value is cannot simply be relied upon. In this book, we sought the origin of the mysterious force of gravity in the vacuum.

Currently, the vacuum is theoretically possessed of matter or energy, and it is recognized that it is not completely empty due to the presence of dark matter and dark energy. We have newly interpreted the light propagation and gravity under the premise that the vacuum is made of a medium with the density much higher than that of matter we know and has near zero energy (zero point energy). Matter is nothing but the energy added to the solid vacuum, and the solid vacuum expands or distorted as much as the energy, yielding stress in the solid vacuum. Based on the physical discoveries and interpretations, it was discussed that the change in the stress of the solid vacuum will be the source of gravity. Also, we reinterpreted cosmological phenomena based on the premise that the movement of matter is made in the form of wave by exchanging energy with the solid vacuum.

I. Two mysterious things: light and gravity

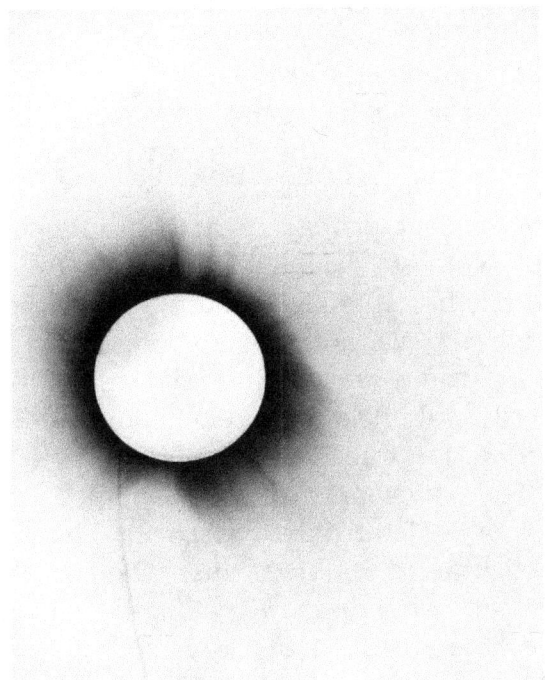

Photograph of the total eclipse by Eddington on May 29, 1919. The results of Eddington's observations were to test Einstein's theory of general relativity that light deflects in the gravitational field. Many newspapers, including the New York Times at the time, gave such enormous repercussions as to feature the results on the front page.

1.1. Light needs a medium for its travel?

Just as the old saying, "It is obvious that it looks like fire," light comes from a fire directly in front of us or far away from a distance of several dozen light years. Light is a electromagnetic wave, as is well known. As is often the case, light is representing the whole electromagnetic waves. What is an electromagnetic wave? It is a term that combines an electric wave and a magnetic wave that propagate as the intensity of the electric field (E) and magnetic field (B) changes vertically with the intensity fluctuating when light passes through the vacuum or a medium. In addition to visible light, radio waves, infrared rays, ultraviolet rays, and x-rays all belong to electromagnetic waves. The difference is due to the difference in the wavelength. The wavelength of visible light is near 500 nanometers (nm, 10^{-9} m) (see Figure 2). The energy of electromagnetic waves is large when the wavelength is short and small when the wavelength is long. The wavelength of electromagnetic waves used for radios or TVs is several meters, and x-rays for medical use belong to the high energy side.

Denied virtual vacuum medium - Ether

Light has been almost completely theorized in the field of optics, and based on the theory, many optical devices and equipments have been developed and used in everyday life. The most distinctive feature of

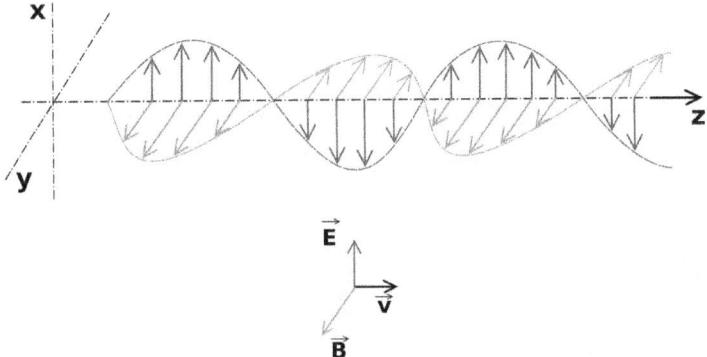

Figure 1. The electric field (E) and the magnetic field (B) of light.

light is the fact that it propagates without any medium, unlike sonic or other types of waves. The speed of light in the vacuum is constant at ~ 300,000 km per second.[2] Because of the nature of light being perfectly straightly propagate in the vacuum, in the 1600s Newton[3] argued that light would be composed of particles and had been accepted as such for a considerable period of time. But since Huygens[4] el al. proved the wave theory of light, light was gradually accepted as a wave. In 1801 Young[5]

[2] The speed of a wave is given as the product of the wavelength (λ) and the frequency per unit time (f). The constant speed of light means that the shorter the wavelength, the greater the frequency.

[3] Isaac Newton (12.1642 - 3.1727), physicist and mathematician of the United Kingdom,

[4] Christiaan Huygens (4.1629 - 7.1695), Dutch mathematician, physicist and astronomer.

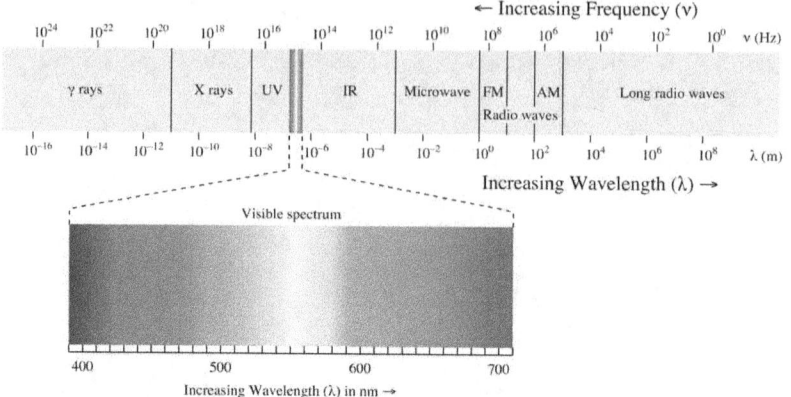

Figure 2. The wavelength of visible light ranges from 400 to 700 nanometers (wikipedia.org).

observed the interference of light in a double-slit interference experiment.[3] Subsequently, the wave theory of light replaced the particle theory.[4]

Scientists who believed that light is a wave in the Newtonian era needed some medium in which light could wiggle and thought of "Ether" to be the medium. Ether is a hypothetical fluid that enables the fluctuation of light. In 1887, to demonstrate that Ether was really exist, Michelson[*6] and Edward Morley performed the famous "Michelson-Molly experiment".[5]

---

*5) Thomas Young (6.1773 - 5.1829) British physician, physicist, physiologist and linguist.
*6) Albert Abraham Michelson (12.1852 - 5.1931) Polish-American physicist. The Michelson-Molly experiment, which denied the existence of Ether, made Michelson to be the first American to win the Nobel Prize in Physics 1907.

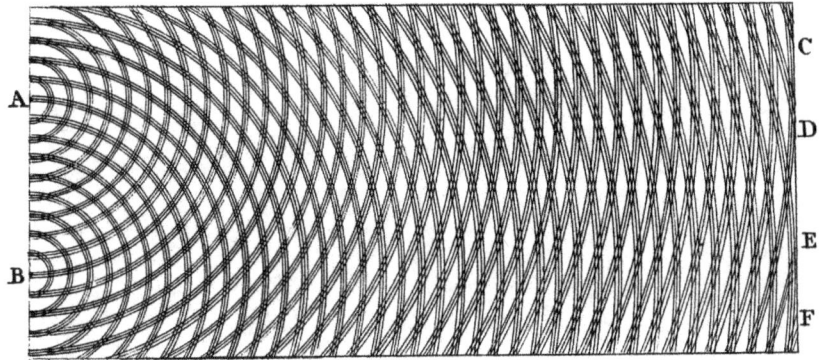

Figure 3. A sketch showing the diffraction of a wave by a double slit of Thomas Young.

For this experiment, Michelson worked with Molly on a sophisticated optical device as shown in Figure 4 to observe the interference expected to appear due to the difference in the speed of light under the influence of Ether.

When Earth rotates around the Sun and the flow of "Ether" should be felt differently in the parallel and in the perpendicular direction to the flow of Ether. If two beams of light move the same distance but in the two different directions, the speed of light of the two beams would be different, and it was expected that the interference pattern of light would appear. However, unlike the expectation, such a interference was not observed. Though more sophisticated experiments had been performed since then, the results were the same.[6] There was no such a thing like "Ether" that could mediate the propagation of

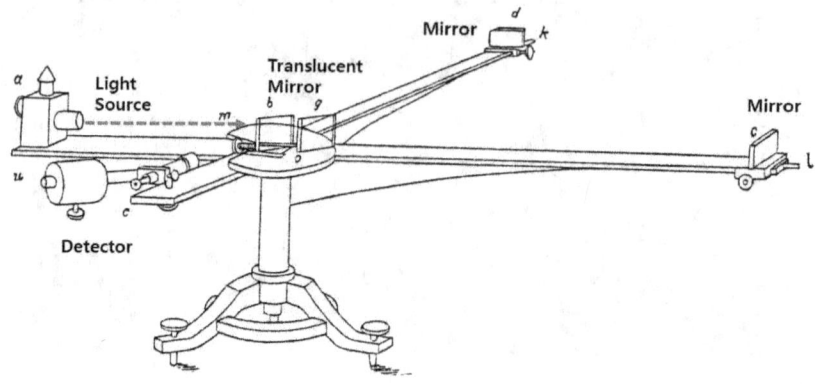

Figure 4. Schematic of the interferometer developed by Michelson in 1881 (modified from the image of wikipedia.org).

light. The experiments that attempted to confirm the existence of Ether rather denied it.

As a result, light could travel in the vacuum without any influence of the vacuum itself, and was recognized as a dual nature with the characteristics of wave and particle as observed in the latter half of the 20th century (the particle characteristics of light, its quantization of energy, contributed greatly to the birth of quantum mechanics, the essence of modern physics). It is not a granular-like particle, but light has the characteristics of a particle in the sense that the energy of light is quantized. It is different from the particles of the Newtonian age.

1.2. Distortion of time - Special Relativity

Based on the fact that there is no medium to sustain light waves like 'Ether' as disproved in the Michelson-Molly experiment and that light travels by itself through the vacuum and the speed of light is constant regardless of the movement of the light source (the same velocity is measured), Einstein[*7)] published the theory of special relativity (Special Relativity) in 1905(But Einstein claimed that the Michelson-Molly experiment had nothing to do with the theory).[7] This theory applies only to inertial systems (non-accelerated coordinate systems) and assumes that the laws of physics in inertial systems do not change. This theory was derived to explain the difference between Newtonian mechanics and electromagnetism in the regime of particles moving at near the speed of light at that time. From this theory, so-called Einstein's mass-energy equivalence was derived, in which mass (m) is compatible with energy (E) via the speed of light (c): $E = mc^2$. This equivalence equation made Special Relativity very solid by fully explaining the relationship between the mass difference and energy released before and after atomic nuclear fission or fusion, and the space in which light travels has no medium such as "Ether". The confidence that nothing is in the vacuum became more and more solid.

[*7)] Albert Einstein (3.1879 – 4.1955) German-born theoretical physicist. The 1921 Nobel Prize in Physics was awarded to him for his theory on the photoelectric effect.

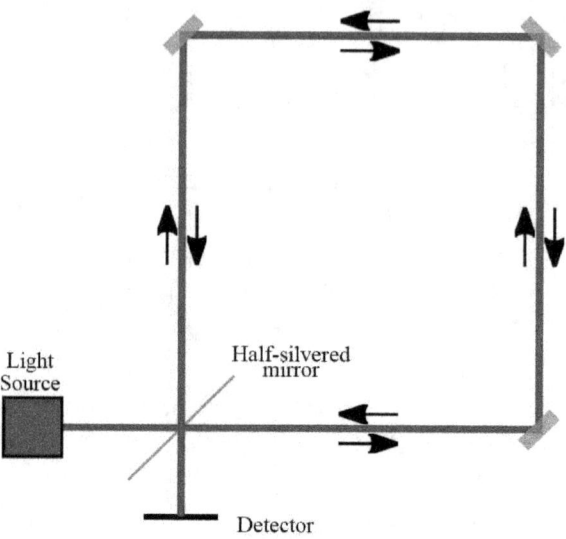

Figure 5. Schematic of Sagnac's light interference test equipment. As the device rotates, the light beam in the rotating direction meets the beam in the opposite direction of rotation at the exit to yield interference.

Sagnac's Experiment

The Michelson-Molly experiment denied the existence of "Ether", confirming that the speed of light in the vacuum is invariant and became the basis of Special Relativity. But Sagnac's[8] experiment was once known to prove the existence of Ether. In 1913, Sagnac experimentally showed the interference of light when a light beam is divided into two parts on a rotating platform as shown in Figure 5. The phase difference

*8) Georges Sagnac (10.1869 - 2.1928) French physicist.

of the interfering light beams is proportional to the rotational angular velocity of the platform.[8] This is the Sagnac effect. Lau[*9] theoretically predicted this phenomenon in advance: if the speed of light and the speed of rotation are constant, there is a time difference of light traveling in opposite directions and interference occurs.[9] This phenomenon, of course, was also consistent with Special Relativity and the existence of stationary Ether, but the effect of Ether expected from the Michelson-Molly experiment was not confirmed. Since then, more sophisticated theoretical considerations have denied the existence of Ether, and light has moved through the vacuum and its speed has become more and more constant.

Theory of length contraction and time dilation

Special Relativity explains phenomena that cannot be felt in our daily life, such as length contraction, time dilation, and relativistic mass. This is fundamentally due to the constancy of the speed of light in the vacuum. To maintain this invariance, time expands and length shrinks.

Length contraction is a phenomenon that length is measured less by a stationary observer when an object is moving rather than in rest.[10] It is also known as the Lorentz contraction. Usually it can be

*9) Max Theodor Felix von Laue (10.1879 – 4.1960). German physicist. He was awarded the Nobel Prize in Physics in 1914 for discovering the diffraction phenomenon of x-rays by solid crystal structures.

recognized only in the direction of travel, when the speed of an object approach the speed of light. This length contraction was originally proposed to compensate for the hypothesis of Ether,[11] the presence of which had been denied in the Michelson-Molly experiment. However, Einstein explained that length contraction is only compatible with his theory of special relativity regardless of the existence of a fluid like Ether, and drastically changed the concept of time and space at that time.[12] This idea of Einstein later has developed into Minkowski's four-dimensional spacetime concept.*10)

Time dilation is a phenomenon in which time is observed to be dilated when observing in a certain inertial system another inertial system having a relative velocity. In a given inertial system, an observer in rest and a moving observer perceive each other's clocks to be late. This is because they move relative to each other no matter who moves. In common sense, if time of a moving object is slow, the object will feel that time of the outside world is faster. Relativity predicts the opposite. This seems to be a paradox. It is like a situation that a person looks another distant person smaller than him or her, and the distant person looks him or her smaller than the person far away. In this sense, time dilation is not a paradox,[13] and this characteristics of time having

---

*10) Hermann Minkowski (6.1864 - 1.1909). German mathematician. He provided his student Einstein his theory of time and space which enabled to understand General Relativity in four dimensional spacetime.

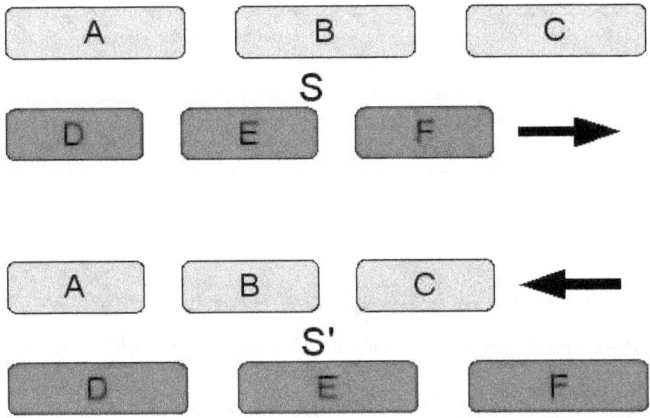

Figure 6. Length contraction: Assume three bars A, B, C lying in a stationary state S and in a motion state S'. When the left end of A and D is in the same position in the horizontal direction, the length of the bar is measured differently. In S, the distance between the left end of A and the right end of C is larger than the interval between D and F. The opposite is true for S'. (Arrows indicate the direction of motion. wikipedia.org).

reciprocity or symmetry, time dilation, is well understood from the relativistic point of view.[14] Time dilation occurs even in environments where the strength of the gravitational field is different or when an object is accelerating. A watch in a high gravitational field is slower than a watch in a weak field. An atomic clock sent to space goes slower than a clock in Earth, or a GPS (Global Positioning System) clock goes faster.[15] However, as the case with the

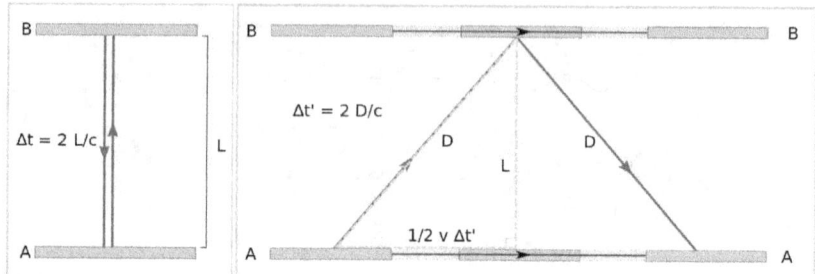

Figure 7. Left: When A and B are stationary, the time taken by the optical signal coming from A and returning from B to B is 2L/c. When A moves from left to right, the optical signal is generated from A at time t' = 0, B receives this signal at time t' = D/c, and A receives the reflected signal at time t' = 2D/c (wikipedia.org).

twin paradox*11) the reciprocity or symmetry of time is not established if not for inertial systems.

Length contraction and time dilation can be expressed by the following equations respectively,

$$\text{length contraction: } L = \sqrt{1 - \frac{v^2}{c^2}} = \frac{L_0}{\gamma} \quad \text{--- } (1.1)$$

and

---

*11) If one of the twin brothers is on Earth and the other travels on a spaceship. Then one brother becomes younger than the other, when the two meets again after the space trip. This paradox can be attributed to the fact that the brother who traveled felt acceleration in the departure, arrival, direction changes. It is time dilation in the situation of acceleration, which is not symmetrical.

time dilation: $\Delta t' = \dfrac{\Delta t}{\sqrt{1-\dfrac{v^2}{c^2}}} = \gamma \Delta t$ --- (1.2),

where γ is called the Lorentz factor[16] and v and c are the velocity of the object and the speed of light, respectively. The length L_0 and the time change Δt at rest contracts and expands to L and $\Delta t'$, respectively, due to the relative movement. Eqs. (1.1) and (1.2) are combined to

$$L_0 \Delta t = L \Delta t' \quad \text{--- (1.3).}$$

The product of length and time is constant regardless of length contraction and time dilation. Therefore, time dilation is substantially the same as length contract and both sides of a coin in Special Relativity.

In Special Relativity, "mass" has two meanings. One is the same constant amount for all observers in all coordinate systems, "rest mass (m_0)" or "invariant mass", which is equal to the "rest energy". The other corresponds to "relativistic mass (m)" or the "total energy" of an object, depending on the velocity of the observer. From Einstein's mass–energy equivalence, all masses are interchangeable with energy and converted to the total energy of the object. The total energy including the kinetic energy is expressed as follows via the Lorentz factor,

$$E = \gamma E_0 = \gamma m_0 c^2 \quad \text{--- (1.4).}$$

In Eq. (1.4), γm_0 is the relativistic mass, having the same form as time dilation in Eq. (1.2). Therefore, energy and time can be perceived as the same, but different only in their form. On the other hand, light has zero rest mass but has energy and thus relativistic mass.

As mentioned above, Special Relativity comes from the constancy of the speed of light moving in the vacuum. However, light can bend or change its speed in strong gravitational fields. It is a phenomenon that cannot be interpreted only by the theory. In the vacuum where there is nothing, light can move by alternating the electric field and magnetic field as shown in Figure 1, but how does gravity occur and affect light through the vacuum of nothing? This is a more mysterious phenomenon than the propagation of light.

1.3. Gravity, the very unknown force

The fundamental forces of Nature are classified into four kinds.[17] They are the strong force, electromagnetic force, weak force, and gravitational force. The strong force, also called the strong interaction or strong nuclear force, is related to the force of nucleus. It holds quarks*[12] within baryons,

---

*12) Quarks are the elementary particles and are known as the fundamental substance of matter. They are not directly observed, only indirectly recognized from the behavior of baryons. Quarks are the only elementary particles in the

such as protons and neutrons. Neutrons and protons are attracted due to this force to form nuclei. Usually the mass of protons and neutrons comes from the energy of the strong force. Quarks contribute only about 1% to the total mass.[18] The weak force, also called the weak interaction or weak nuclear force, is associated with radioactive decays in atoms and plays an important role in the nuclear fission process. The electromagnetic force, familiar to us, is a force that occurs between charged particles. An electromagnetic field such as an electric field, a magnetic field, or an electromagnetic wave, is handled within the category of this force. In high energy particle physics, the weak force and electromagnetic force are integrated into the electroweak force.[19]

The weakest force that rules Nature - gravity

The relative strength compared to the strong force is 10^{-2} for the electromagnetic force, 10^{-9} for the weak force, 10^{-38} for the gravitational force, being very weak compared to the other three kinds of forces. Gravity is a very familiar term of physics together with electromagnetism, but it is also the most mysterious and difficult one regarding to its nature in modern physics. Theories on gravity have been developed since the work of "Mathematical principles of natural philosophy" by Newton in 1687, through

---

standard model of particle physics, which react to the four forces. Their charge has a fractional value. There are six kinds of quarks: up, down, strange, charm, top and bottom quarks.

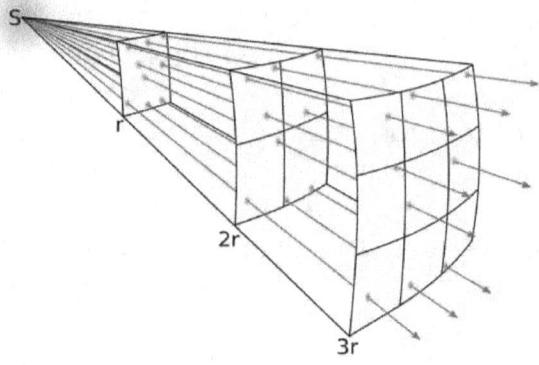

Figure 8. S is a light source. The amount of light energy is constant regardless of the distance. Therefore, the amount of light energy per unit area is inversely proportional to the square of the distance from the source, since the surface area of the sphere is proportional to the square of the radius. Similarly, the gravitational strength of matter is inversely proportional to the square of the distance, since the origin of gravity is matter wave, as will be discussed in Chapter 2.

Einstein's theory of general relativity (General Relativity) to quantum gravity*13).20

---

*13) Quantum gravity is a theory of quantum mechanics and belongs to a field of theoretical physics to explain gravity. Quantum effects cannot be neglected near the strong gravitational field of a massive object. Although "gravitons" are introduced in the theory to interpret the concept of gravity, it has limitations that no meaningful predictions are possible.

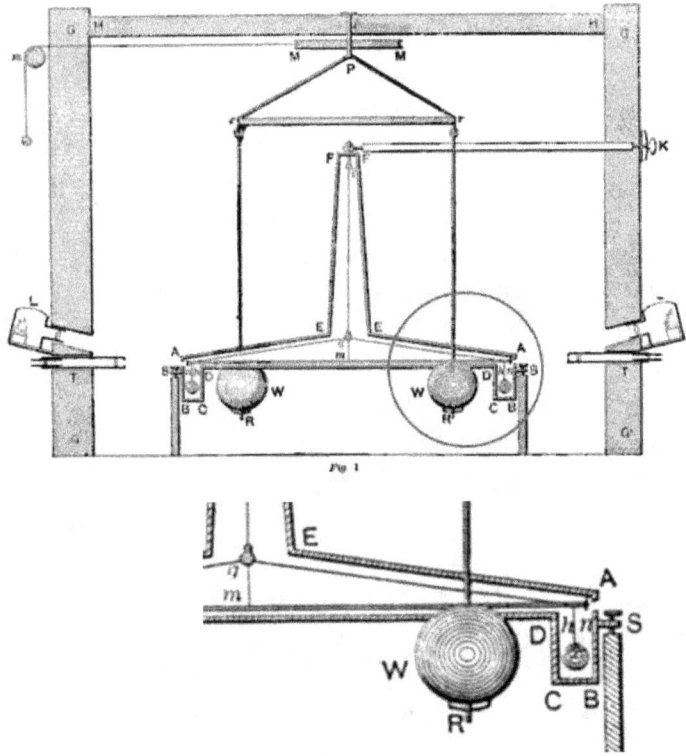

Figure 9. A cross-sectional view of the Cavendish torsional balance device for measuring the gravitational constant. The cross-section below is an enlargement of the circle in the upper one. The hanging large sphere W is rotated by the outer sheave and moves to the side of the small sphere x.

According to Newton's theory of gravity, the gravitational force (F) acting between two objects with mass m_1 and m_2 is given as

$$F = \frac{Gm_1 m_2}{r^2} \quad\text{---}\quad (1.5).$$

F is inversely proportional to the square of the distance r between the two objects (see Figure 8). Eq. (1.5) does not have any background theories and is a principle of experience. In Eq. (1.5) G is the gravitational constant. It was first experimentally verified by Cavendish[*14] in 1797~1798 using a torsion balance shown in Figure 9.[21] At that time, G was measured as 6.74×10^{-11} $m^3 kg^{-1} s^{-1}$, being only 1% different from the currently certified value 6.674×10^{-11} $m^3 kg^{-1} s^{-1}$. As of 2014, it was reported to be 6.67191×10^{-11} $m^3 kg^{-1} s^{-1}$.[22] Figure 10 shows the evolution of the measured values. It is seen that they converge to the current official value.

Gravity is the force we feel most closely in our daily life. Standing upright or jumping high must also consume energy to overcome the potential energy of Earth's gravitational field. Having the more kinetic energy in the radial direction than the gravitational potential energy on Earth one can escape Earth. This velocity is called the escape velocity and is calculated as 11.186 km/sec.[23]

Newton's theory of gravity also explains the cosmic phenomena such as Kepler's laws. For example, Kepler's[*15] third law can be derived from the theory.

---

*14) Henry Cavendish (10.1731 ‑ 2.1810) British natural philosopher and scientist. He is famous for the discovery of hydrogen, which he himself called "the combustible air."
*15) Johannes Kepler (12.1571 ‑ 11.1630) German mathematician,

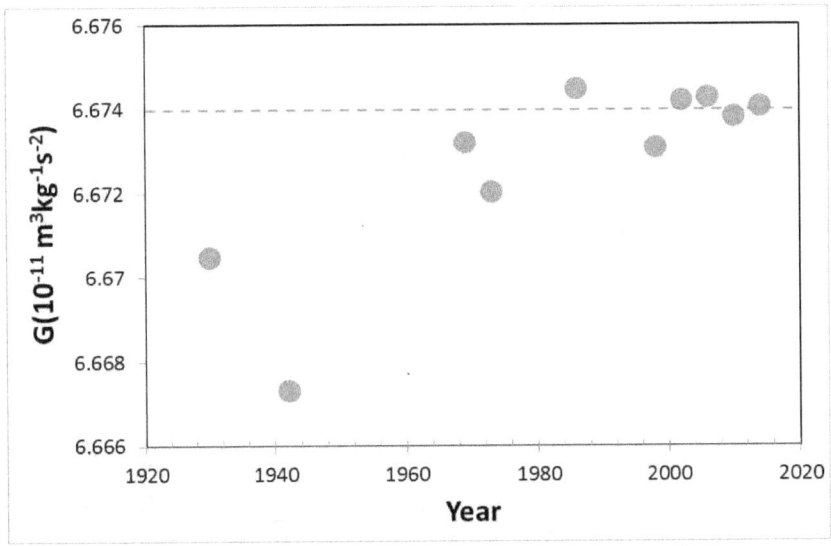

Figure 10. Changes in the measured gravitational constant (the dotted line is the officially approved value).

This law states that the square of the orbital period of a solar planet is directly proportional to the cube of the semi-major axis of the orbit, and it represents the relationship between the distance to the Sun and the orbital period. Kepler referred to this law as "the music of the planet" for its neat "mathematical representation". This law has since been known as the law of harmony.[24] According to Newton's gravity law, if the centrifugal and gravitational force are the same, the following equation holds for a circular orbital.

---

astronomer and astrologer. Observing and analyzing the motion of the planets, he found Kepler's laws with his own name. These laws form the basis of Newton's theory of gravity.

$$mr\omega^2 = mr\left(\frac{2\pi}{T}\right)^2 = \frac{Gm_1 m_2}{r^2} \quad \text{---} \quad (1.6),$$

where ω is the angular velocity and T is the orbital period, respectively. By rearranging the above equation we have

$$T^2 = \left(\frac{4\pi^2}{GM}\right)r^3 \quad \text{---} \quad (1.7).$$

For an elliptical orbital, it is calculated based on the center of gravity. In this case, the radius r is replaced by the semi-major axis a and the solar mass M is replaced by $M+m$ (m, the planetary mass). In other words,

$$\frac{a^3}{T^2} = \frac{G(M+m)}{4\pi^2} \approx \frac{GM}{4\pi^2} \quad \text{---} \quad (1.8).$$

As $m \ll M$, m is ignored. The prediction of Eq. (1.8) is 7×10^{-6} [AU^3/day^2].[16] Table 1 shows the calculated ones from a and T of the solar planets. It is seen that all the planets are well following the prediction.

What is gravity?

Where does the gravitational force come from and what is its real nature? Even Newton, who published

---

[16) AU is 149,597,870,700m, the distance between the Sun and Earth.

Table 1. GM/$4\pi^2$ of the solar planets
(Wolfram|Alpha Knowledgebase 2018)

planet	semi-major axis a [AU]	orbital period T [days]	$a^3/T^2 \times 10^{-6}$ [AU^3/day^2]
Mercury	0.38710	87.9693	7.496
Venus	0.72333	224.7008	7.496
Earth	1	365.2564	7.496
Mars	1.52366	686.9796	7.495
Jupiter	5.20336	4332.8201	7.504
Saturn	9.53707	10775.5990	7.498
Uranus	19.19130	30687.1530	7.506
Neptune	30.06900	60190.0300	7.504

the theory of gravity, wrote in his letter to Bentley[*17)] in 1692: "It is inconceivable, that inanimate brute matter, should, without the mediation of something else, which is not material, operate upon and affect other matter without mutual contact..."[25] The origin of gravity has yet to be clarified. Some researchers present theories on its origin related to entropy.[26] It is called entropic gravity and describes gravity as a force by entropy but it is not a fundamental force. This theory is based on string theory,[*18)] black hole

---

[*17)] Richard Bentley (1.1662 – 7.1742) British classical scholar, critic and theologian.
[*18)] In physics, particles are points whose size is zero, but string theory is based on a string having a certain size.

physics, and quantum information theory, and explains that gravity is a phenomenon caused by quantum entanglement[*19] of small bits of spacetime information. Thus entropic gravity sustains by the second law of thermodynamics, where entropy increases always with time. Entropic gravity provides a framework for describing Modified Newtonian Dynamics (MND). In MND, the gravitational force decreases in inverse proportion to the distance from the center of mass, not the square of the distance, at a threshold of about 1.2×10^{-10} m/s^2. This value is very small, only 12×10^{-12} of the gravitational strength on Earth's surface, which is equivalent to 36 hours for an object to fall from 1 m above the surface. Scientists who believe this theory agree with Newton's theory of gravity and General Relativity in spacetime distortion, but has not yet been proved.

[*19] Quantum entanglement states that the quantum states of individual particles cannot independently be described when multiple particles or pairs of particles are created or interacted (even though they are far apart) and should be described as a whole system. The position, momentum, spin, and polarization of entangled particles are connected to each other. For example, when a pair of particles is generated with zero total spin, if one particle has a spin in the clockwise direction, then the spin of the other particle is in the opposite direction. Measuring the state of a particle is like applying energy to the particle, and if it changes its original quantum state, this measurement is done on the entire entangled particle system for entangled particles. Thus, one particle in a entangled pair immediately sees that another particle is being measured, even though there is no means of exchanging information with each other.

1.4. General Relativity with no gravity

Since the origin of gravity is unclear, Einstein called the gravitational force a phenomenon that is originated from the properties of time and space, which are distorted by mass and energy, rather than a certain force. This characteristic was formulated in his theory of general relativity and published in 191 5.[27] It was ten years after Special Relativity had been published.

Einstein's spacetime distortion

The mathematical equations of General Relativity are called Einstein's field equations. They consist of a total of 10 differential equations, and represented by

$$R_{\mu\nu} - \frac{1}{2} R g_{\mu\nu} + (\Lambda g_{\mu\nu}) = \frac{8\pi G}{c^4} T_{\mu\nu} \quad \text{---} \quad (1.9),$$

where $R_{\mu\nu}$ is the Ricci curvature tensor and describes the distortion of the space. $g_{\mu\nu}$ is the metric (shortest distance) tensor in a 4-dimensional spacetime, which is a combination of space and time called Minkowski spacetime. $T_{\mu\nu}$ is an energy-momentum tensor describing the density and flow rate of energy and momentum. Λ in the parentheses on the left-hand side of Eq. (1.9) is the cosmological constant and was absent in the publication of 1915, but was included in the modified version of 1916.[28] In this book, we will not deal with Einstein's field equations in more detail.

Figure 11. Distortion of spacetime around Earth according to General Relativity (Image: @ NASA).

They are conceptually difficult and complex to deal within the scope of this book. However, the equations can be visualized as a virtual image as shown in Figure 11. Due to Earth's mass (the right side of Eq. (1.9)) spacetime around it is distorted (the left side of Eq. (1.9)).

The field equations converge to Newton's theory of gravity in the regime where gravity is weak or the velocity of an object is insignificant compared to the speed of light. However, if the gravitational field is strong or the velocity approach the speed of light, it explains well the phenomena which the Newton theory cannot. For example, Einstein predicted quantitatively that light is bent when passing through a strong gravitational field, as proved by Eddington. In 1919,

Figure 12. Photograph of the total
solar eclipse taken by Eddington
on May 29, 1919.

Eddington[20] measured the angle of deflection of
starlight on the island of Principe, west Africa, and
found that it bent about 1.6"[21] on passing the Sun.[29]
Figure 12 is a picture of the total solar eclipse taken
at that time.[22] It was very close to the prediction of

---

[20] Arthur Stanley Eddington (12.1882 – 11.1944) British
    astronomer.
[21] 1" (arcsecond) = 1/3600 degrees.
[22] The results of Eddington's observations were to prove
    General Relativity, the impact of which was so enormous that

General Relativity, 1.75". According to Einstein, the reason for the deflection of light is that spacetime near the gravitational field is distorted as shown in Figure 11. According to mass-energy equivalence ($E = mc^2$), the energy of light is converted to mass, and starlight can be attracted to the Sun in its gravitational field. So the deflection is also predicted by Newton's theory of gravity, but it is only half the predicted by Einstein's field equations. In addition, General Relativity has been recognized as the best theory of the cosmology by clearly explaining phenomena such as the precession of Mercury,[*23] the gravitational redshift.[*24] etc.[30]

Einstein' s equations and quantum mechanics

It is known to be very difficult to obtain exact solutions of Einstein's field equations in Eq. (1.9). What is known as an exact solution is an approximate expression in a strict sense. The Schwarzschild metric, derived for a weak gravitational field condition, is typical.[31] It was published in 1916, the year following the publication of Einstein's equations, and was named after the German mathematician

---

many newspapers at that time, including the New York Times, featured on the front page.

[*23] A phenomenon that the perihelion (the point closest to the Sun) moves while Mercury orbiting the Sun around an elliptical orbit. See Section 2.8 for details.

[*24] A phenomenon in which the wavelength of light from a light source placed in a strong gravitational field becomes reddish to an observer in a weak gravitational field.

Schwarzschild.[*25)] This is a good description of the orbital motion of Earth in a weak gravitational field, but the problem of singularity in a strong gravitational field was raised, which became the mathematical birthplace of black holes.[*26)]

Einstein mentioned that the universe would not shrink if there is some energy not observed in the universe, although it is supposed to shrink due to gravity. He has introduced a constant, the cosmological constant Λ, for this invisible energy as in Eq. (1.9). Λ, a kind of vacuum energy, cannot be measured, but it must exist to prevent the shrinking of the universe caused by gravity. This cosmological constant was not attracted any attention at first, but it has been regraded as dark energy since Hubble's observations that the universe would acceleratedly expand. The presence of unknown dark energy can allegedly explain the accelerated expansion of the universe predicted from the redshift phenomenon of light coming from distant galaxies.[*27)] If dark energy promotes the expansion of the universe, then dark matter suppresses it.

Dark matter[*28)] is a concept introduced in order to explain the observation that the orbital velocity of stars orbiting far away from the center of a galaxy

---

[*25)] Karl Schwarzschild (10.1873 – 5.1916), first sought a solution to the field equations in the same year when General Relativity was published.

[*26)] They are called the Schwarzschild black holes and we will discusses on the paradox of black holes in detail in section 3.7.

[*27)] dark energy, refer to section 4.3.

[*28)] dark matter, refer to section 4.2.

faster than predicted by Newton's theory gravity or General Relativity. If the velocity is faster than the predicted by the law of gravity, the stars should be repelled from the orbit due to the centrifugal force. Therefore, some additional mass in the gravitational field must be added to the observed mass to make up the prediction. This additional mass should be so-called "dark matter". The theory of entropic gravity explains that dark matter, so called it is invisible, is actually a product of the quantum effect, and can be regarded as a form of positive dark energy given to the vacuum in the ground state.[32] Dark matter has not yet been found in these predictions or interpretations, and so far it is assumed to be merely a mean to cover the problems of the theories of gravity.

General Relativity, which comprises Newton's theory of gravity and Special Relativity, explains various cosmological phenomena and predicts strange phenomena such as black holes, dark matter, and dark energy. All of these are new physical phenomena that we have not experienced yet. Whether these predictions will all be proved is something we should wait, but there are still bigger homework: It is the fact that General Relativity and quantum mechanics are incompatible.[33] In quantum field theory, gravity cannot be described by imaginary gravitons, as contrarily as electromagnetism is described by virtual photon exchanges.[34] In the macroscopical world, it is consistent with the classical mechanics or General Relativity, but not in the very microscopic region of the Planck length.*[29) New theories on quantum mechanics or gravity or fusion theories are seemed to be required.

Loop quantum gravity (LQG) theory is a kind of fusion theory, that merges quantum mechanics and General Relativity. Starting from the theory of relativity, the properties of quantum mechanics are added to integrate gravity into the theoretical framework that deals with three other fundamental forces, the strong force, electromagnetic force, and weak force. A similar theory, string theory, adds gravity to quantum field theory. Like energy and momentum in quantum mechanics, LQG is one of the key factors for the development of quantum theory based on Einstein's spacetime geometry. Quantum mechanics quantizes or particularizes time and space, as with photons or atomic energy levels. There is a minimum distance to quantize the space. This is the Planck length and there is no shorter length. The space itself resembles the atomic structure.[35] New theories, including this theory, have broadened the concept of vacuum, but have not yet developed into a big theory that covers quantum mechanics and General Relativity. The emergence of "Theory of Everything" appear to be still a long way to go.

---

*29) One of the fundamental units of nature derived from the Planck constant h, the gravitational constant G, and the speed of light c by Max Planck in 1899. It is given as $l_p = \sqrt{hG/2\pi c^3}$, and the exact value is $1.616229(38) \times 10^{-35}$ m. The Planck constant is a constant that connects the energy of photon to the frequency. The photon energy is given as the product of the Planck constant and the frequency. The value is $\times 10^{-34}$ J·s.

II. Origin of gravity

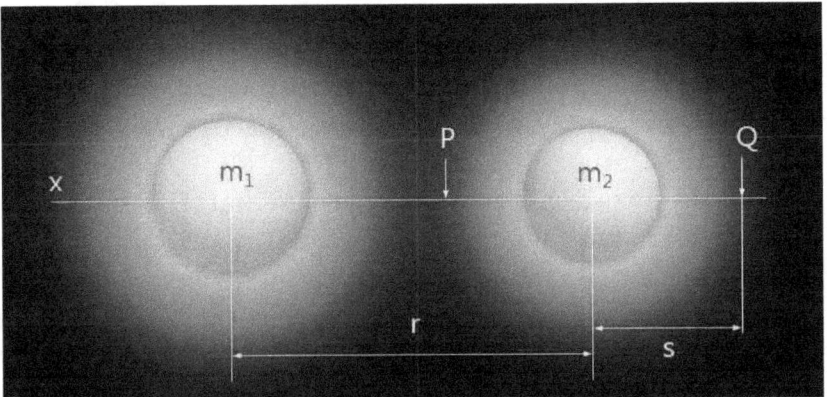

An object in the solid vacuum will distort the medium as much as the mass of the object, and the intensity of distortion decreases away from the center of the object in a point-symmetric manner. The distortion of the solid vacuum around an spherical object m_2 becomes asymmetric due to the nearby presence of m_1. The intensity of distortion is higher at point P than at Q, though the two points are at the same distance from the center of m_2, and m_2 moves to the side (P) where the distortion is more severe. This is the law of inertia. Furthermore, as m_2 moves toward m_1, the difference in the distortion becomes stronger, so m_2 accelerates toward m_1. This is the origin of gravity.

2.1. Universe filled with a very hard medium

Until quite a long time ago, the vacuum was literally nothing. The question of whether the vacuum is really nothing seems to have been raised when the wave theory of light was dominant over the particle theory for the nature of light. There was a thinking that there must be a medium for light to propagate in the vacuum, if light should be a wave. But it was denied by the Michelson-Molly experiment. In the process of appreciating the duality of light after that the energy of light had been accepted to be quantized, the discussion on the medium for the light propagation has disappeared. Einstein provided Special Relativity based on the invariance of the speed of light in the vacuum of nothing. But the argument that the vacuum can have energy was also provided by Einstein. According to him, as mentioned in Chapter 1, there must be some necessary energy in order for the universe not to shrink inevitably due to gravity. Einstein called this "the cosmological constant" and included it in his field equations of 1916. The existence of energy in the vacuum is indeed confirmed by the observation of the Casimir effect, spontaneous emission, and the Lamb shift.

Is there energy in the vacuum?

As shown in Figure 13, imaginary photons in a pair of parallel plates with a very short distance can only

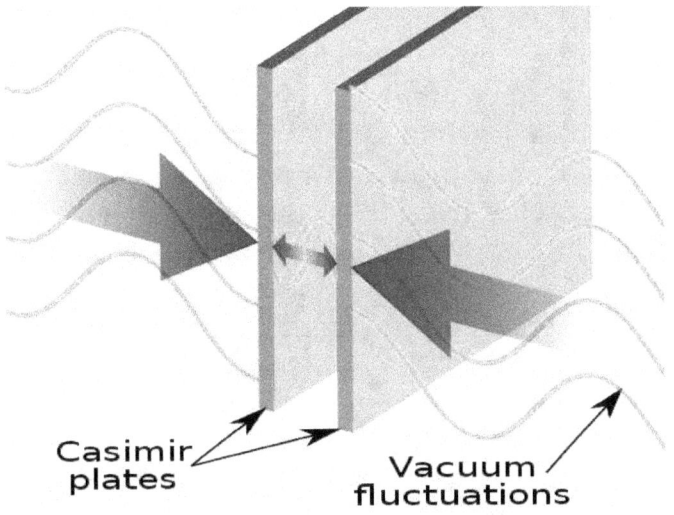

Casimir
plates

Vacuum
fluctuations

Figure 13. The Casimir effect (wikipedia.org).

exist in the form of standing wave,*30) so the number of photons inside of the plates is lower than that of outside the plates in the vacuum. This difference gives pressure on the plates. This is the Casimir effect.36 Although this force was very small, the measurement was done between two plates only few nanometer apart, and it turned out that quantum fluctuations*31) indeed exist in which the vacuum oscillates in terms of energy according to Heisenberg's uncertainty principle.*32)

---

*30) A wave that does not move but vibrates only in place.
*31) It is a transient energy change at a point in space according to Heisenberg's uncertainty principle. This process generates a particle-antiparticle (e.g, electron-positron) pair.
*32) A principle discovered by German physicist Werner Karl

Spontaneous emission is the process of emitting a quantum in the form of photon while a quantum mechanical system (atom, molecule, corpuscular) transits from a excited state to the ground state. Spontaneous emission is the source of all kinds of light. Lasers are devices that use stimulated emission, but driven by spontaneous emission. Spontaneous emission is a process of quantum mechanics and is explained in terms of the zero-point energy of the electromagnetic field.[37] In the Jaynes-Cummings model,[38] a theory of quantum optics, spontaneous emission is explained by the interaction of an atom with two quantum states with the optically quantized vacuum. The emission rate of photons can then be controlled via the boundary conditions of the ambient vacuum.

The Lamb shift, on the other hand, is the subtle difference in the energy levels of a hydrogen atom having the same energy in the Dirac equation.[*33] The

---

Heisenberg (12.1901 - 2.1976, won the Nobel Prize in Physics in 1932). When measuring two mutually related physical quantities simultaneously, there is a physical limit of the accuracy. For example, the position and momentum of a particle cannot be measured at the same time. The higher the accuracy of the position, the lower the accuracy of the momentum.

[*33] A theory published by Paul Dirac (9.1902 - 10.1984) in 1928, which is a relativistic quantum wave equation describing Fermi particles with a spin of ½ like electrons. Dirac described the behavior of electrons with this equation and predicted the existence of positron as the antiparticle based on the mirror symmetry of the equation. In 1932, an US physicist Anderson (Cahl David Anderson, 9.1905 - 1.1991) experimentally verified the equation by discovering positrons. Dirac won the Nobel

interaction between quantum fluctuations and the hydrogen electron is the cause of the Lamb shift. This effect was confirmed in the Lamb-Retherford experiment[*34)] on the hydrogen microwave spectrum in 1947.[39]

The vacuum energy represented by the cosmological constant, the zero-point energy, is a very small amount of about 10^{-9} J/m^3.[40] On the contrary, according to quantum electrodynamics(QED[*35)])[41] and stochastic electrodynamics(SED[*36)])[42] the vacuum energy can be unbelievably as high as 10^{113}J/m^3.[43]

Prize in Physics with Schrödinger in 1933 for his discovery of this "new form of atomic theory". Anderson also won the Nobel Prize in Physics in 1936 for discovering positrons.

[*34)] Willis Lamb (7.1913 – 5.2008, US physicist) received the Nobel Prize in Physics in 1955 for the Lamb-Retherford experiment.

[*35)] QED in particle physics is a relativistic quantum field theory of electrodynamics. QED explains how light interacts with matter, and it is the theory that first fully unifies quantum mechanics and Special Relativity. QED mathematically describes all the phenomena involved in interacting charge particles by exchanging photons. It is a perturbation theory of the electromagnetic quantum vacuum. It can calculate the abnormal magnetic moments of electrons and the Lamb shift of the hydrogen energy level very precisely.

[*36)] At zero absolute temperature, the electromagnetic energy in the vacuum is depicted as a randomly fluctuating zero-point energy. In this random oscillation, particle motions are very nonlinear, disorderly, and protruding. In SED, quantum properties of waves and particles are seen as protruding effects due to this nonlinear matter-vacuum interaction. Therefore, quantum mechanics is a special case included in the nonlinear theory of SED, and non-local signal transmissions are possible and the uncertainty principle is broken. Inertia is also one of these nonlinear laws in SED.

When this value is converted to mass from Einstein's mass-energy equivalence, $E = mc^2$, the mass density of the vacuum becomes $5.16×10^{94}$ g/cm^3. This value is called the Planck density[44] and is very large compared to the density of $1.73×10^{15}$ g/cm^3 of a proton converted from the proton volume $9.66×10^{-39}$ cm^3 and the mass $1.672×10^{-27}$ kg.

New vacuum paradigm

In conclusion, the vacuum is not a state of "nothing". It has energy or mass. However, the difference in the amount of vacuum energy experimentally measured, observed or theoretically predicted is too large. This difference is called the cosmological constant problem or "vacuum catastrophe" in physics. There are still no discoveries or satisfactory theories to overcome this gap. It is impossible to measure such a enormous vacuum energy.

Let's imagine here. The very high vacuum density should mean that the vacuum is made of a very dense and solid medium and light propagates through this medium and has very little energy (zero-point energy) at the ground state. In this case, the cosmological constant problem is solved and the vacuum catastrophe disappears. All the cosmological phenomena we see and feel occur within the solid vacuum, and this solid vacuum has the ground state (minimum) energy as much as the cosmological constant. The density of the solid vacuum is the absolute density of the vacuum that can never be

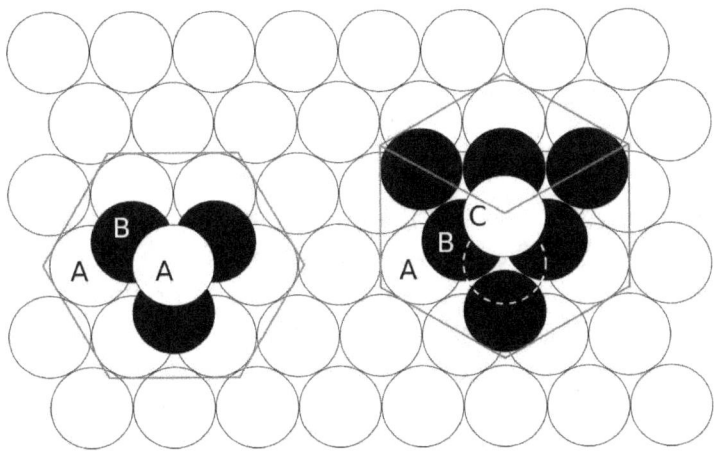

Figure 14. The hcp (left) and fcc lattice. Each lattice is outlined by straight lines. Alphabetic characters indicate the position of constituent particles. A-B-A-B in hcp, and A-B-C-A-B-C in fcc (wikipedia.org).

converted to energy. This is our paradigm of the vacuum and has the following specific premises.

Firstly, the vacuum is filled with a very high density (neutral) medium*37) and is not converted to energy. This solid vacuum assumes to form a lattice with neutral virtual particles. Such a lattice may have the most dense face centered cubic (fcc) or hexagonal close packed (hcp) lattice. As shown in Figure 14, these lattices have the highest packing factor (74%)

---

*37) The medium, not matter, is distinguished from matter in the universe.

and are most stable in terms of energy when spherical particles of the same size regularly fill the space. In order to deform or break this structures, more energy must be applied than any other ordered structures.

Secondly, when (enough) energy is input to the cold solid vacuum, electrons and positrons are generated (as particle-antiparticle pairs are generated via quantum fluctuations), and these positrons are trapped in the vacuum lattice to become protons. Protons and electrons combine to form atoms which in turn constitute molecules. That is, when energy is input in the cold solid vacuum, ordinary matter is created.

Thirdly, when an electron is captured by a proton and keeping a certain distance from the center of the proton to form a hydrogen atom, it distorts the nearby solid vacuum (Of course, the proton itself has already distorted the solid vacuum). As shown in Figure 15, the electron of a hydrogen atom is around the proton at a certain distance (being present while oscillating or rotating). Protons and electrons push or expand (strain) the solid vacuum. This is like making an air bubble in a material made of rubber by blowing high-pressure gas. This is matter in our vacuum paradigm. When the temperature of matter increases, that is, when the energy is input to the cold solid vacuum, the volume expands as much as the input energy.

In the regime of our vacuum paradigm, everything

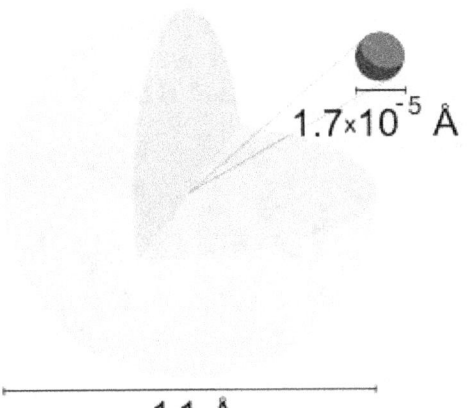

1.7×10^{-5} Å

1.1 Å

Figure 15. Hydrogen atom model
(wikipedia.org).

we see and feel is a composite of energy and the cold
solid vacuum. That is, "matter = the cold solid
vacuum + energy". If the structure of this composite
or the bonding force between the constituent particles
changes, the mass of the composite must changes.
Before and after a nuclear fission or fusion process,
the total amount of charge or the number of
constituent neutrons/protons do not change, but a
mass deficiency occurs. This is because the distortion
intensity of the solid vacuum changes. Referring to
Figure 16, when ^{235}U decays into ^{92}Kr and ^{141}Ba as a
result of the collision with a free neutron, additional
two neutrons are generated. The neutron number is
1(colliding neutron) + 235(^{235}U) = 236 before collision
changes to 92(^{92}Kr) + 141(^{141}Ba) + 3 (generated

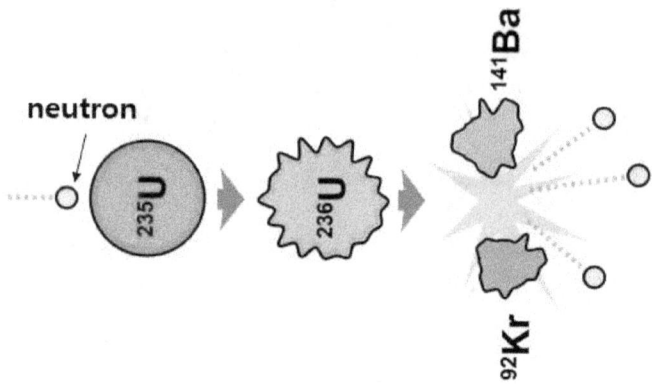

Figure 16. Nuclear fission (wikipedia.org).

neutrons) = 236. Namely, the neutron number is the same in the process of nuclear fission. Mass deficiency were caused by changes in the binding energy of the constituent particles before and after the reaction.

Actually, various elements in the universe are from hydrogen atoms through nuclear fusion. The proton and electron in an hydrogen atom merge to form a neutron, which is a building block for helium and heavier atoms.*38)

Helium with two neutrons is very stable. The two neutrons of helium are in the state of lower energy

---

*38) Figure 43 shows the process of nuclear fusion of helium from hydrogen in the Sun, where protons are deprived of positrons to be neutrons. These positrons combine with the electrons of hydrogen and disappears as gamma rays. As a result, neutrons are generated from protons and electrons.

than that of two pairs of proton and electron. The amount of energy stored in the solid vacuum will be very low, compared with that of 4 separate hydrogen atoms. In the theory of particle physics, neutrons and protons are not elementary particles, and there are elementary particles such as quarks. Most of these elementary particles are coming into being at very high energy levels and have very short lifetimes and thus cannot be the components of the cold solid vacuum of little or no energy. Since there is no energy-free particle, even a elementary unit of matter, only the neutral lattice point can be assumed as the elementary particle in the cold solid vacuum. When positrons are captured in the neutral lattices, they become protons. When electrons are trapped around protons, they constitute hydrogen atoms. When the electrons of hydrogen atoms are fused with the protons, neutrons with high energy are generated. When the energy of neutrons is exhausted, leaving only the neutral vacuum lattice free of energy. This is the energy cycle in our vacuum paradigm.

2.2. Light is trembling of the solid vacuum

The speed of all electromagnetic waves including light in matter is slower than in the vacuum. So, the refractive index[*39)] is always larger than 1.[45] As shown in Figure 1, light moves in the vacuum with an

*39) refractive index, the speed of light in the vacuum divided by that in matter.

electric field and a magnetic field crossing each other vertically in a wave form. The difference in the traveling speed means that the mechanical properties of the cold solid vacuum is different from those of matter. In regime of our vacuum paradigm, light does not propagate in the space of nothing, but it moves through the highly dense medium, the solid vacuum. Since the propagation speed through a medium is determined by the properties of the medium itself, it can be understood that light in an inertial frame with no gravitational field moves 300,000 km/sec, and is no more or less than that. It is of the course, particles (or waves) faster than light cannot exist because of the nature of this medium. So how does one know the properties of the solid vacuum that allows the speed of light to be 300,000 km/sec? First, let's look into the case of sound waves. They propagate not only in air or liquid, but also in solids.

Propagation of sound waves in solids

Sound waves in solids are due to volume deformation (compression) and shear deformation, and are divided into pressure waves (longitudinal waves, P waves) and shear waves (transverse waves, S waves). The speed of sound waves through a homogeneous three-dimensional solid is given respectively as [46]

$$\text{P waves: } v_p = \sqrt{\frac{3K+4G}{3\rho}} = \sqrt{\frac{Y(1-\nu)}{\rho(1+\nu)(1-2\nu)}} \quad \text{---} \quad (2.1),$$

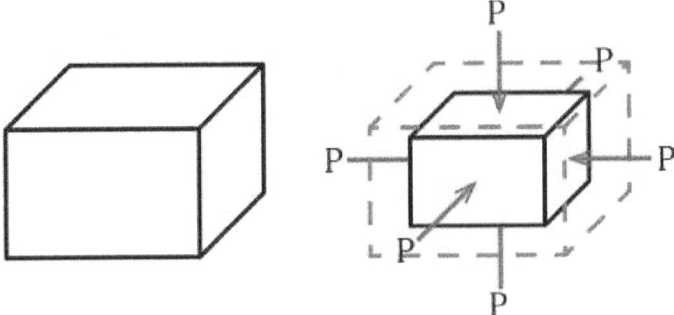

Figure 17. The bulk modulus K is the ratio of the volume deformation rate (dV/V) to the external homogeneous pressure P. $K = -\dfrac{dP}{dV/V}$ (wikipedia.org).

$$\text{S waves: } v_s = \sqrt{\frac{G}{\rho}} \quad \text{---} \quad (2.2),$$

where K is the bulk modulus (see Figure 17), G is the shear modulus (see Figure 18), Y is Young's modulus, ρ is the mass density, and v is Poisson's ratio.[*40], given by $Y = 3K(1-2v)$, respectively. Eq. (2.1) and Eq. (2.2), v_P depends both on the volume and shear deformation characteristics of the material, but v_S is only a function of the shear modulus.

[*40] The ratio between the strain in the tensile direction and in its perpendicular direction. Most materials have a value of 0.2 to 0.3. Concrete has a value 0.1 to 0.2, and cork ~ 0. A perfect incompressible material with a volume unchanged against external pressure has the theoretical maximum of 0.5.

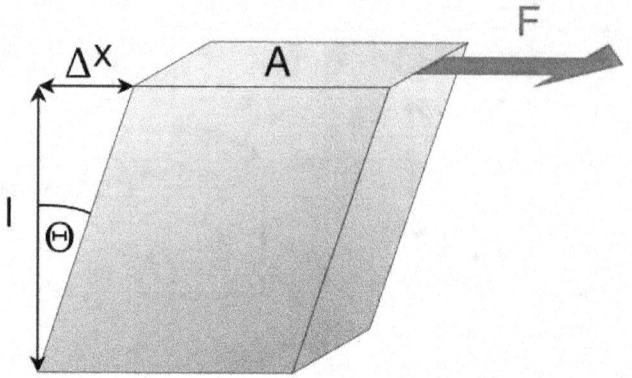

Figure 18. The shear modulus G is the ratio of the shear stress (τ) to the shear deformation(γ).
$$G \equiv \frac{\tau}{\gamma} = \frac{F/A}{\Delta x/l} \quad \text{(wikipedia.org)}.$$

Usually P waves are faster than S waves. Therefore, when an earthquake occurs, the vertical vibration first occurs, and the lateral vibration follows. For alloys, K = 170 GPa, G = 80 GPa, ρ = 7,700 kg/m³. By inserting these values into Eqs (2.1) and (2.2), v_P is calculated to be 6,000 m/s[47] (measured one is 5,930 m/s)[48] and v_S is 3,200 m/s. A sound wave through a long bar can be regarded as a one-dimensional solid sound wave, and v_P for this case is given as

$$v_p = \sqrt{\frac{Y}{\rho}} \quad \text{---} \quad (2.3).$$

The speed of P waves through a long bar is always lower than that through a homogeneous three-

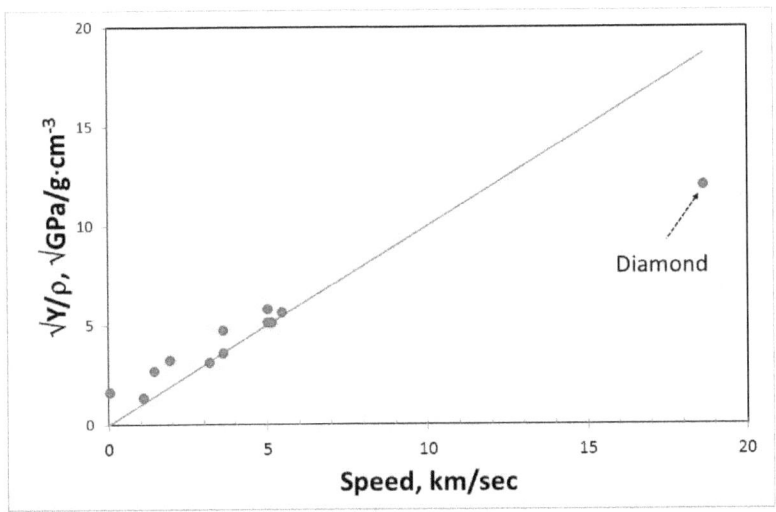

Figure 19. The speed of sound waves depending on the materials properties. The solid line is the case where the calculation by Eq. (2.3) and the measured value are the same.

dimensional solid, and the difference depends on Poisson's ratio of the material.

Figure 19 and Table 2 show the correlation between the speed of sound waves and the properties of the materials. Most solids satisfy Eq. (2.3), but the calculated velocity is somewhat higher than the actual one. Diamond is known to be the fastest material in sound waves. As shown in Figure 19, its calculated value is much lower than that of the measured one. The speed of light is ~25,000 times faster than that of diamond of ~12 km/sec.

Table 2. Measured and calculated speeds of sound waves from Eq. (2.3) in various solids

	speed km/sec	density g/cm^3	GPa Modulus	$\sqrt{Y/\rho}$
Diamond	12	3.5	1220	18.67
Pyres glass	5.64	2.23	67	5.481
Steel(long)	5.79	7.82	200	5.057
Steel(trans)	3.1	7.82	80	3.198
iron	5.13	7.89	210	5.159
Aluminum	5.1	2.7	69	5.055
Brass	4.7	8.48	112	3.634
Copper	3.56	8.79	117	3.648
Gold	3.24	19.29	74	1.959
Lucite	2.68	1.18	2.5	1.456
Lead	1.322	11.35	13.8	1.103
Rubber	1.6	1	0.0025	0.05

Similarity of sound waves and light

As shown in Figure 1, light is a S wave rather than a P wave. Then, Eq. (2.2) can be applied. By inserting the speed of light in Eq. (2.2), it is possible to determine one of the properties of the cold solid vacuum. If the density of the solid vacuum is the Planck density of 5.16×10^{94} g/cm^3, the shear modulus of the cold solid vacuum is calculated to be 4.64×10^{100} GPa. This may be called the Planck shear modulus. Of course, since the mediums for sound wave and light are different, this value may not indicate a physical property of the cold solid vacuum. However, from Fig. 19, it is seen that, if the propagation speed is large, the density and the elastic modulus of solids are large, and then we assume that the physical properties of the cold solid vacuum through which light propagates would be very different from those of our materials.

Here we note a similarity between Einstein's mass-energy equivalence and the equation of the speed of sound waves in solids. The speed of light c by the equivalence is given as

$$c = \sqrt{\frac{E}{m}} = \sqrt{\frac{G_V}{\rho_V}} \quad \text{--- (2.4)}.$$

ρ is the density of the solid vacuum in m/V and $G_V = E/V$, the shear modulus of the cold solid vacuum, if this equation and (2.2) are identical. Thus, mass-energy equivalence can be regarded an equation representing the propagation of light through the cold solid vacuum.

Here, we may raise a question, whether there are only shear waves due to charge fluctuations, namely electromagnetic waves. Are there any non-charge vibrations in the cold solid vacuum? Neutrinos move at the same speed of light, but matter does not interact with neutrinos. In this sense, neutrinos can be regarded as P waves in the cold solid vacuum, while light is one of S waves. Since these two kinds of waves are the same in the speed, we can obtain the relationship between the bulk modulus and the shear modulus by equating Eq. (2.1) and Eq. (2.2). Namely,

$$3K = -G \; \text{---} \; (2.5).$$

The bulk modulus has a negative value!! The negativity of the bulk modulus is thermodynamically impossible, as is evident in Figure 17. When an external pressure is applied, the volume must be reduced. Although the bulk modulus should have a positive value, a negative elastic modulus may appear. A negative value is measured in a foamed plastic material such as styrofoam already deformed.[49] The bulk modulus can also be negative when there is a volume change in phase transformations, as with the change in the crystal structure.[50] It is a very special case. As will be explained in Section 4.3, it is interesting to note that the universe acceleratedly expand because dark energy acts as "negative pressure". When the space expands thermodynamically, the energy is reduced because it works on the outside. However, since dark energy represented by the cosmological constant increases due to a volume expansion, the pressure becomes negative. In this case, the bulk modulus can be negative. Does the cold solid vacuum have a negative value in the bulk modulus? We may agree that the bulk modulus of the cold solid vacuum is negative, if we think that the cold solid vacuum is incompressible because it contains no energy and it will swell up when energy is added to it in our vacuum paradigm.

2.3. The starry night of Gogh

There would be almost no one who do not know the Dutch great impressionist painter, Vincent van Gogh (3.1853 - 7.1890). Among his 900 paintings left, "The Starry Night (Figure 20)" is regarded as the masterpiece in Western cultural history.[51] This painting depicts the sky and village that he saw just before sunrise seen in his room at the Saint-Remy-de-Provence hospital where he stayed till his death. Stars in the dawn sky look strange. The night sky is filled with something, and a star cluster seems to flow along the ripples between the stars. Did Gogh know the secret of the vacuum and gravity? The picture give us a strong impression that the sky is full of something and is distorted around stars.

Distortion of the solid vacuum

Our vacuum is full of a very dense medium like shown in the painting of Gogh. Matter in the universe is nothing but energy trapped in this solid vacuum in the form of atoms and molecules. Atoms themselves deform the surrounding solid vacuum as much as their mass or energy. Figure 21 shows an imaginary stress field around a sphere in the cold solid vacuum. This stress field appears to be very similar to the concept that Einstein's spacetime is distorted around matter in 4-dimensional spacetime as shown in Figure 11 and is similar to the image of stars in Figure 20. The concept of our cold solid vacuum and matter

Figure 20. "The Starry Night" painted by Gogh in June 1889. Since 1941, it is a permanent collection of the Museum of Modern Art in New York.

(atoms and molecules) makes it easy to understand the distortion of Einstein's spacetime. It is not a strange spacetime distortion, but an interaction of the cold solid vacuum and the energy of matter that should be considered in search of the origin of gravity.

If the strength of spacetime distortion in the regime of General Relativity reflects the magnitude of

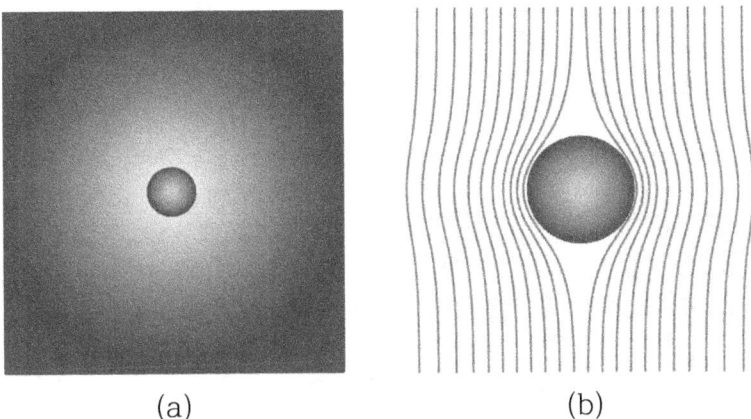

(a)   (b)

Figure 21. (a) A stress field around a spherical object in the solid vacuum. The closer to the sphere, the greater the compressive stress. (b) An imaginary stress distribution around a spherical object in the x-axis direction.

gravitational force, how can such a force be interpreted in the regime of our vacuum paradigm? Assume that a spherical object with mass m_1 is inserted in a stress-free space (here, for convenience, a two-dimensional space) as shown in Figure 21(b). We can assume that the density of an inserted object is nearly zero because there is a huge difference in the density between the solid vacuum (say the Planck density) and matter we actually know in the universe. To easily understand, we can think matter is like pressurized gas bubbles in a huge rubber material. If things are inserted into the stress- and strain-free (cold) solid vacuum, it will be pushed to be deformed and a stress (or deformation) field will develop as shown in Figure 21. Tensile stress will be formed parallel to the surface of the inserted material and compressive stress will develope in the radial direction from the center of the object. It is a concept similar to Einstein's spacetime distortion.

To what extent will the inserted material deform the nearby solid vacuum? Calculations based on the conventional elastic theory show that the radial deformation around the solid vacuum is inversely proportional to the square of the distance from the center,[41] when the solid vacuum is swollen as shown in Figure 21. Compared with the Newtonian gravity equation of Eq. (1.5), it is conjectured that the deformation of the solid vacuum is the source of the gravitational force. If the spherical object is a solid, stress due to this deformation will form inside the object, and there will be a singularity where stress becomes infinite as shown in Figure 22. If the center is an empty space or a liquid, this singularity will not form.

Is the distortion of the cold solid vacuum the origin of gravity?

Suppose that another spherical object with mass m_2 is inserted away from the center of the first object m_1 by the distance r (distance between the center of the first object and that of the second) as shown in

---

[41) Three equilibrium equations and twelve stress-strain equations are required to calculate based on the elastic theory. Detailed calculations are omitted in this book.

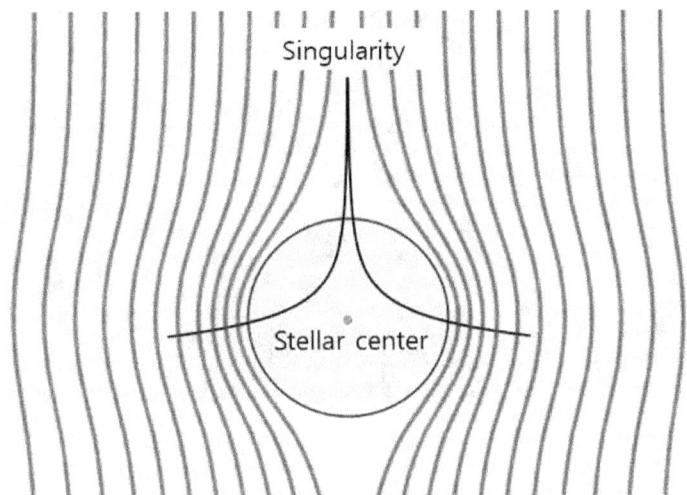

Figure 22. Imaginary stress distribution inside a spherical object placed in the solid vacuum.

Figure 23. Let the line connecting the two central points be the x-axis. When an arbitrary point between the two objects on the x-axis is denoted by P, the distortion of the solid vacuum by m_1 and m_2 are overlapped at point P. Let's simplify the discussion and assume that the size of two objects is the same and point P is right in the middle of the two objects. At this point, the (compressive) stress in the x-axis direction will cancel each other and the tensile stress will be added to the direction perpendicular to the x-axis (being the y-axis) to double. Namely, when all the stress between the two objects are added together, the compressive stress in the x-axis direction is canceled out and the stress in the y-axis direction is

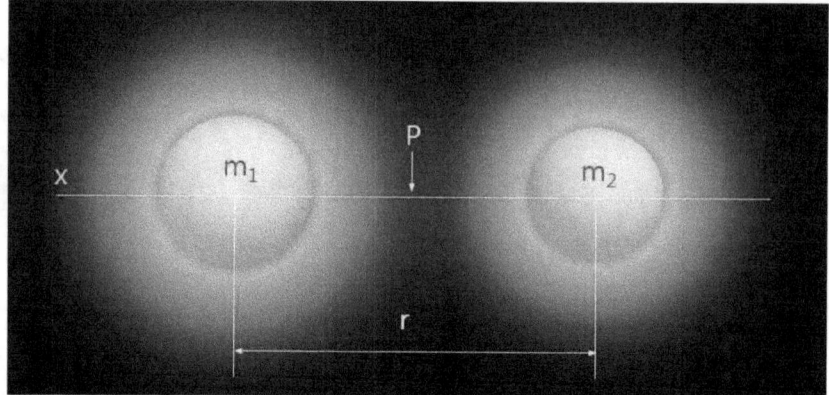

Figure 23. Two spherical objects adjacent to each other.

doubled.*42) When the y-axis is stretched in the elastic theory, m_1 and m_2 are attracted to each other. May this be the origin of gravity. Is the gravitational force known so far the force that arises from the distortion of the solid vacuum? But here comes a question. As the two objects approach, the compressive stress between the two objects will increase to offset the original tensile stress and eventually reach an equilibrium and stop. This is quite different from the nature of gravity we know. How do we solve this contradiction? This will be discussed in Section 2.6.

---

*42) Even if the two objects are different, the compressive stress along the x-axis is partially canceled and the tensile stress along the y-axis is added up.

2.4. Mass: energy stored in the solid vacuum

Mass is a property of an object, a measure of resistance to acceleration due to external forces. It also determines the strength of gravity among objects. So far, mass has been defined in many ways. Among them, what we can most easily access is related to gravity, gravitational mass.[*43] It is simply calculated by weight divided by the gravitational acceleration on Earth. The other, inertial mass, is defined from the relationship between the force F and the acceleration a when an object of mass m is accelerated. That is, $m = F/a$. Inertial mass and gravitational mass are conceptually quite different, but so far no differences have been confirmed experimentally.[52] But classical theories like Newtonian mechanics do not explain why these masses are the same. It is just an experience-based principle. General Relativity begins with the assumption that these two kinds of mass are inevitably the same. This is called the equivalence principle. In this theory gravity is not a force, nor is the object of Newton's action and reaction law (the third Newton's law of motion). It is still a mystery that inertial mass and gravitational mass are the same.[53]

[*43] Gravitational mass is divided into active and passive gravitational mass. The former defines mass as a measure of the strength of the gravitational field created by an object and the latter defines mass as a measure of the strength for an object to react in a gravitational field. In other words, if the magnitude of the gravitational force on the surface of the Earth is F and the magnitude of gravitational acceleration is g, the passive gravitational mass of the object becomes F/g.

Inertial mass and gravitational mass

The fact that inertial mass and gravitational mass are the same is called "Galileo's equivalence principle" or "the weak equivalence principle".[54] This principle applies to free falling objects. When inertial mass is m and gravitational mass is M, if the gravitational force is the only force, according to Newton's 2nd law of motion and the gravity law the acceleration a is given as

$$a = \frac{M}{m}g \ \text{---} \ (2.6).$$

If all objects fall equally in the gravitational field, the ratio of gravitational mass to inertial mass is constant. This phenomenon is called "the universality of free-fall". The first person to test this universality is Galileo. He increased the precision of time by rolling a ball over a nearly frictionless sloping plate. In 1889, Eötvös[*44)] performed more precise experiments using a torsion balance.[55] As of 2008, the precision is at least 10^{-12} and will be within the tolerance range of 10^{-13} to 10^{-18}.[56] On the other hand, the strong equivalence principle, also called the Einstein equivalence principle, is the key of General Relativity. According to it, it is impossible to distinguish between the uniform acceleration and

---

[*44)] Loránd Eötvös (7.1848 – 4.1919) Hungarian physicist. His achievements in research on the weak equivalence principle and gravity has influenced General Relativity.

uniform gravitational field in a sufficiently small spacetime. Therefore, this theory assumes that the force to an object due to gravity is the result of the tendency of the object to move in a straight line.

Are all masses converted to energy?

According to mass-energy equivalence, mass is interchangeable with energy. That is, $m = E/c^2$. In the regime of our vacuum paradigm, energy is stored in the form of distortion of the solid vacuum (or the corresponding compressive stress in matter). Mass is just another expression of the distortion of the cold solid vacuum. Can all these masses be converted to energy? The answer to this can be somewhat inferred from nuclear fusion or fission reactions. As mentioned in Section 2.1, there are no changes in charge and in the number of protons and neutrons before and after the fission reaction (see Uranium fission in Figure 16). In this reaction, the mass deficit is due to the binding energy (potential energy) difference between particles. In our vacuum paradigm, the binding energy comes from the variation of the distortion of the solid vacuum. On the other hand, as shown in Figure 24, it is known that during beta decay,[45] a neutron is transformed into an electron (or positron) and a

---

[45] Beta decay in nuclear physics is a phenomenon in which β -rays (high energy electrons or positrons) and neutrinos come from unstable nuclei. During beta decay, a neutron emits an electron and becomes a proton, or a proton emits a positron and become a neutron.

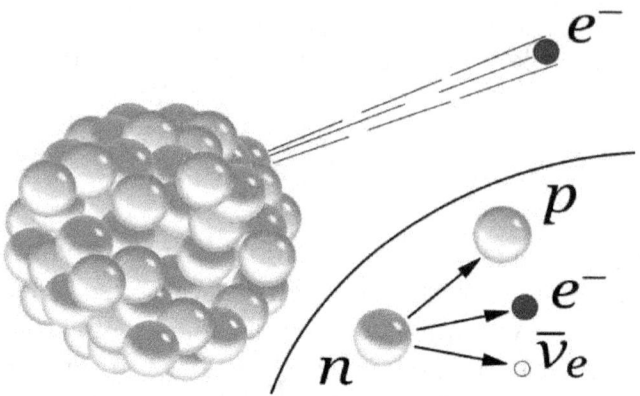

Figure 24. Beta decay. A neutron (n) in an atom emit an electron (e^-) and a neutrino (ν_e) and become a proton (p) (wikipedia.org).

proton in the atom.

We may say that all the mass of a neutron is not converted to energy. Let's however look into the nuclear fusion process that occurs inside a star, as shown exemplarily in Figure 43. This process yields one helium atom out of four hydrogen atoms with some energy released. Neutrons generated in this process are lower in energy than hydrogen atoms in proton–electron pairs. A fission reaction changes a (free) neutron into a proton (+ electron), and a fusion reaction converts a proton (+ electron) into a neutron. Both the reactions emit energy. Neutrons involved in these two processes are different. Neutrons generated by nuclear fusion are lower in energy, and helium, being composed of two neutrons and two protons, is a

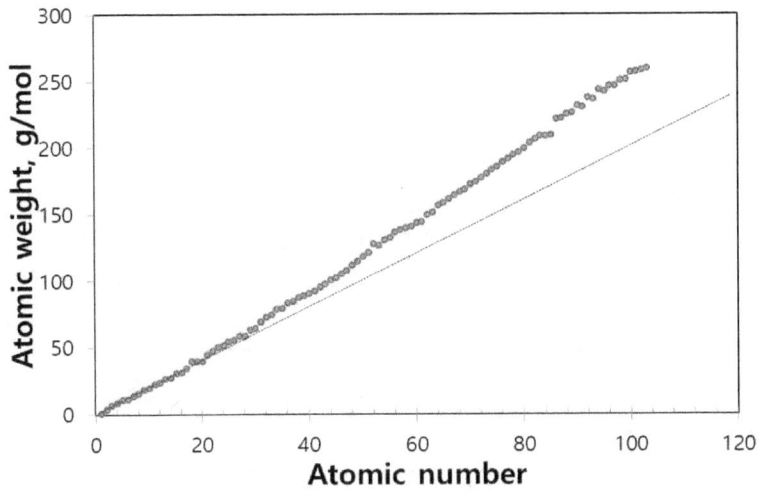

Figure 25. The atomic number vs. the atomic weight. The dotted line is the atomic number × 2.

very stable element. Just as helium is generated by fusion inside of a star, all other elements form inside a star via nucleosynthesis.*46) As shown in Figure 25, the higher the atomic number, the greater the number of neutrons than that of protons. The number of surplus neutrons is known as the neutron excess, and because heavy elements rich in these surplus neutrons are formed in the milieu of super-high temperatures and pressures of giant stars (see Figure 54), these neutrons should have lower energy. Thus, the question of whether all of the mass is converted

*46) Refer to Section 3.5, "Supernova – Explosion of the compressed energy" for details.

into energy can be answered as follows.

"A proton-electron pair transforms into a high energy neutron, then to a stable neutron, and finally to a surplus neutron under the influence of high temperatures and high pressures in the interiors of stars, and eventually loses all its energy and returns to the cold solid vacuum (undistorted vacuum lattice). We can say that mass is all converted to energy (According to a work of SK,[*47] a proton has the minimum lifetime, and eventually transforms to a positron and a neutral pion or an anti-muon and a neutral pion).[57]

The cold solid vacuum is regarded as a medium free of energy or with only a negligible amount of energy of about 10^{-9} J/m^3 (zero-point energy). It is stress- and strain-free when no energy is stored, so "gravity" does not work and thus we do not feel the existence of the solid vacuum. Light also travels in the cold solid vacuum at the constant speed. However, when passing through a distorted region of the solid vacuum near an object, it will deflect towards the object, just as expected from General Relativity that light bends in the gravitational field. Light deflects in a distorted region of the solid vacuum when the speed of one direction (the radial direction toward the object) differs from that of the other direction (the circumferential direction), such as when a car turns

---

[*47] A neutrino observatory in Japan. The observatory was designed to detect high-energy neutrinos to search for proton decay, study solar and atmospheric neutrinos, and keep watch for supernovae in the Milky Way Galaxy.

right or left. This deflection will be calculated in Section 2.7*[48] and compared with the prediction of General Relativity.

In summary, the constitutes of our universe are divided into two kinds: one thing that cannot be converted into energy (the cold solid vacuum) and the other that is interchangeable between energy and matter, and the convertible amount is mass. The presence of mass distorts the energy-free cold solid vacuum. When energy (namely converted mass) is released in the form of electromagnetic waves or neutrinos, the distortion of the solid vacuum is mitigated as much as the released energy. If a star is exhausted by releasing all the energy, surplus neutrons generated in the star make a huge atom. This is a neutron star. Some neutron stars rotate at very high velocities and form very strong magnetic fields.*[49] If even this energy goes away, they will turn into black holes, which will eventually disappear when their remained energy is finally exhausted. This is the process of returning to the energy-free cold solid vacuum. A black hole is never a monster that sucks everything including light due to its huge gravity, but it can sucks everything just because it is an empty region of the cold solid vacuum in the regime of our vacuum paradigm.*[50]

---

*[48] Deflection of light in the distorted solid vacuum.
*[49] Refer to section 3.6 for neutron stars.
*[50] Discussions whether black holes actually exist are given in Section 3.7.

2.5. Matter wave: the essence of the movement of matter

We inferred in Section 2.3 what the origin of gravity in the regime of our vacuum paradigm is. Assuming that the vacuum is an elastomer such as rubber and there is a pressurized air bubble in it, this bubble is regarded as an object in space, and the distortion of the solid vacuum developed around this object shall be the origin of the "gravitational" attraction, as the distortion resembles the gravitational field. It would be more appropriate to assume a foamed polymer in the rubber-like vacuum medium, such as shown in Figure 26, because matter is also a part of the solid vacuum, though matter should have a much lower density than that of the cold solid vacuum. From the stress distribution of the solid vacuum in Figure 23, it is seen that the distortion will be greater between the two objects than outside of the two objects. The reason why the two objects in Figure 23 are attracted by gravity is due to the difference in the distortion of the solid vacuum, and as the distance of the two objects decreases, the difference in the distortion increases and the velocity of approach between the two objects increases. But why should there be a force if there is a difference in the distortion of the solid vacuum? This can be attributed to the interaction between matter and the solid vacuum through energy exchange. Here we would like to set up an important premise regarding to the movement of matter through the solid vacuum.

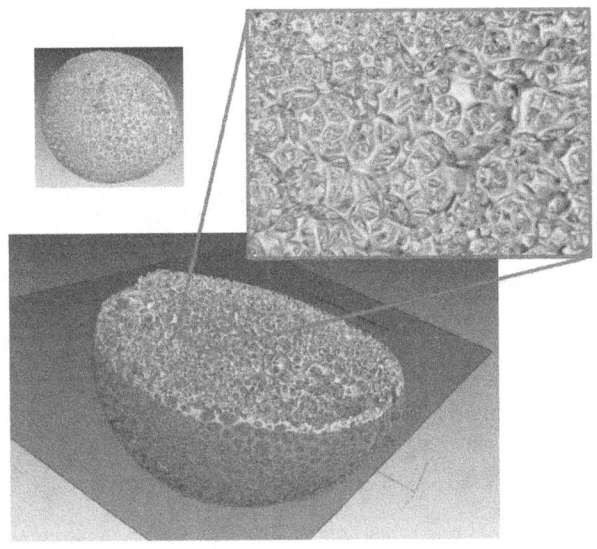

Figure 26. Matter in the solid vacuum is analogous to a foamed polymer (image from wikipedia.org). Matter can have a variety of structures depending on the amount of internal energy and distorts the cold solid vacuum as much as the amount of energy or mass.

Propagation of *vibration* energy of *the* *vacuum* lattice

Matter moves through the solid vacuum in the form of waves (vibrations). Lattice points that make up the solid vacuum only vibrate as much as the energy of matter. There is no movement of the solid vacuum itself, only the vibrational energy is transferred

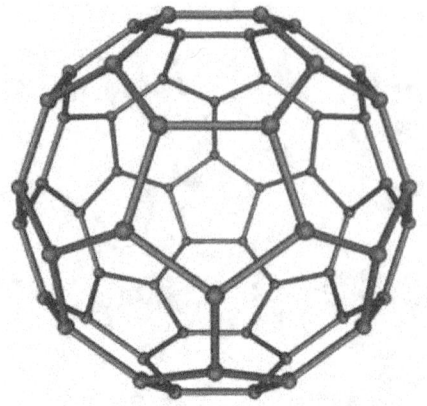

Figure 27. Fullerene shape. A crystal composed of 60-70 carbon atoms in the form of a soccer ball. Diameter is about 1 nm(wikipedia.org).

through the vacuum lattice. This is the very nature of the movement of matter.

There is, of course, the concept of matter wave in physics. Matter wave originally proposed by de Broglie [*51)] in response to the duality of light (wave and particle).[58] The de Broglie wavelength λ is inversely proportional to the momentum p of a particle, as

$$\lambda = \frac{h}{p} = \frac{h}{mv} \quad \text{---} \quad (2.7).$$

[*51)] Louis Victor Pierre Raymond de Broglie (5.1892 – 3.1987). He won the Nobel Prize in Physics in 1929 for his hypothesis on matter wave after confirmed by the experiment of Davidson and Germer in 1927.

where h is Planck's constant. In fact, this hypothesis was confirmed not only for electrons and atoms, but also for polyatomic molecules such as fullerene (C60) in Figure 27, and matter is also accepted to have the duality of particle and wave. In 1927, at Bell Labs,[52] Davidson[53] and Germer[54] experimentally confirmed that electron beams from nickel crystals make diffraction patterns, which was the first to demonstrate the wave character of electrons hypothesized by de Broglie.[59] In 1999, the Vienna scientists passed a fullerene ball through a diffraction grating to find interference,[60] where the wavelength of matter wave was 2.5 pm (pm = 10^{-12} m).

Movement of matter wave and matter

However, matter wave cannot find any correspondence with real physical quantities, elastic/electromagnetic waves. Quantum mechanics interprets this wave based

---

[52] The official name is Nokia Bell Labs, also known as AT & T Bell Labs. It is a research and development company in New Jersey (head office) in the United States owned by Nokia, Finland. It was started as "The Volta Laboratory and Bureau", founded by Alexander Graham Bell at the end of the 19th century. In 1925, Western Electric's research and development became independent of Bell Labs. John Bardeen and two other researchers developed transistors for the first time and received the Nobel Prize in Physics in 1957.

[53] Clinton Davisson (10.1881 - 2.1958) US physicist. The Davidson-Germer experiment proved the wave character of electrons and he won the Nobel Prize in Physics in 1937 for this experiment.

[54] Lester Germer (10.1896 - 10.1971) US physicist.

on a probability function. How does a physical body move in the solid vacuum in the form of waves? In the regime of our vacuum paradigm, matter is energy stored in the cold solid vacuum. If matter should move in the form of a wave through the solid vacuum, it is an energy propagation. Matter is just the vibrational energy of the vacuum lattice, and this vibration will be the source of matter wave.

Imagine a case where a hydrogen atom, the simplest element, moves through the solid vacuum. This hydrogen atom occupies a lattice point of the solid vacuum and an electron is rotating or oscillating at a certain distance around a proton.[*55] We further imagine that the electron is absorbed instantaneously into the proton of the former hydrogen atom to become a neutron and disappear in the lattice point, then the former hydrogen atom appears to have disappeared. At the same time a neutron is generated at a nearby new location, where the electron and proton are re-emerged as a new hydrogen atom. By repeating this process, we recognize that a hydrogen atom is moving. This behavior of energy movement can be described by a wave function of vibration, which corresponds to the behavior of matter wave for a single atom in current physics. Not only a hydrogen atom but also massive objects shall move in the same way. In this case, however, only atoms on the surface exchange energy with the solid vacuum, while the

---

[*55] It cannot be confirmed that it is actually spinning. It is described quantitatively as the probability of being within a certain distance around the proton, see Figure 15.

internal materials exchanges energy between atoms. For this process, the wave function resulting from the energy propagation of each atom can be combined into a single wave packet,[*56] to be recognized as the movement of matter in current physics.[61]

How is then matter wave different from electromagnetic waves? Matter wave and electromagnetic waves have something in common in the sense that they carry energy. Unlike electromagnetic waves, matter wave is emitted by the vibration of lattice point of the solid vacuum itself and this vibration travels through the solid vacuum as matter wave. A hydrogen molecule, for example, is seated at some vacuum lattice points, and vibrate over the center of mass with the frequency corresponding to the energy of the hydrogen molecule, so that the surrounding solid vacuum is distorted in a three dimensionally symmetric manner, when it vibrates only in one place. The movement of the hydrogen molecule means changes in the vibration site, a subsequent procedure of expansion-contraction of the lattice points. On the other hand, electromagnetic waves do not involve such a procedure. Therefore, light has no rest mass. It is as if sound waves are due to the relative vibration of the constituent elements of a solid, and sound waves themselves have no mass. According to de Brodie, the

---

*56) In physics, wave packets are referred to as wave bundles. Two or more waves combine to form a local burst of wave. A number of waves with different properties can be combined into one wave packet, where these individual waves combine locally but cancel each other out in other regions.

product of the propagation velocity (w) of matter wave and the velocity (v) of a particle is equal to the square of the velocity of light (c).[62] In other words,

$$w = \frac{c^2}{v} \quad \text{---} \quad (2.8).$$

Since v is always lower than c, w is greater than c, according to Eq. (2.8). If $v = 0$, $w = \infty$, and as v increases, w decreases and approaches c. In this sense, matter wave is similar to tachyon*[57] that moves faster than light.[63]

In general, the speed of a wave is given as the product of frequency and wavelength, so the frequency (f_m) of matter wave is obtained by combining Eq. (2.7) and Eq. (2.8). Namely,

$$f_m = \frac{mc^2}{h} \quad \text{---} \quad (2.9).$$

In Eq. (2.9), it is apparent that the mass of matter is nothing but the vibration of the solid vacuum. It is nothing but one of our premises that mass and energy are just vibration of the (imaginary) lattice points of the cold solid vacuum.

Let's look into Eq. (2.7) again. According to this equation, if the velocity of matter is zero, the wavelength of matter wave is infinite. It does not

---

*[57] A fictitious particle that moves faster than light. Most physicists deny the presence of tachyon because particles that are faster than light violate current laws of physics.

make any wave in the cold solid vacuum. As the velocity increases, the wavelength becomes shorter. The shorter the wavelength, the greater the momentum. Since the velocity of matter is usually very low relative to the speed of light, the mass of Eq. (2.7) is the rest mass, and thus the velocity of matter is inversely proportional to the wavelength. If the movement of matter is substantially the result of the interaction of matter with the solid vacuum, the wavelength shall be a measure of the distortion of the solid vacuum in the direction of the movement. This length becomes shorter as the mass of matter and the velocity increases. It is the Doppler effect[*58] stating that the distortion of the cold solid vacuum changes with the velocity of matter. Therefore, the wavelength of matter wave is expressed as the product of the distortion of the solid vacuum due to the presence of matter and the additional distortion due to the movement of matter. This is a very important clue to access the fact that inertial mass and gravitational mass are the same and thus to the origin of gravity.

2.6. Gravity cannot exist on its own

We have argued in Section 2.3 that gravity is due to the distortion of the solid vacuum by matter. In

---

[*58] A phenomenon in which frequencies and wavelengths change with the relative speed of a wave source and an observer, discovered by Christian Doppler (11.1803 ‒ 3.1853, Austrian physicist).

General Relativity, gravity is related to the distortion of spacetime. However, General Relativity does not tell how the distorted spacetime makes an attractive force between separated physical bodies. As shown in Figure 23, when two objects are in close proximity, an attractive force due to this distortion develops between the two objects, but if there is no energy exchange between the objects and the solid vacuum, the two objects will stop at the equilibrium position. The attraction force will then be is zero. However, the actual attractive force is stronger as the two objects get closer inversely proportional to the square of distance. To solve this contradiction, it is assumed that inertial mass and gravitational mass have the same origin. Mass is also assumed to have its origin in the distortion of the solid vacuum as much as its corresponding energy.

Distortion of the solid vacuum due to a linear motion

Let's first consider an inertial movement. When a swan moves over the surface of water as shown in Figure 28, its motion will form a concentric wave that continuously spreads out in all directions. If the swan moves at a constant velocity, the wavelength will be shorter in the forward direction and longer in the backward direction. Similarly, consider that an object moves through the solid vacuum at a constant velocity. The object will move at the velocity unless any external forces are applied. This is Newton's law of inertia. The distortion of the solid vacuum yields a

Figure 28. A Doppler effect appears when a swan moves over the water (wikipedia.org).

Doppler effect as with the case of a swan moving on the water surface at a constant velocity. Because energy is stored in the solid vacuum, energy is not consumed by an inertial motion. As the Doppler effect of Figure 29 shows, when an object moves at a constant velocity, the distortion of the solid vacuum increases in the forward direction and is relaxed in the opposite direction. Also, all the solid vacuum in the direction perpendicular to the traveling direction will be distorted. Since the distortion of the solid vacuum is energy (mass), the kinetic energy due to the movement is stored as a kind of distortion energy in the solid vacuum. In other words, an object moves when there is a difference in the distortion of the

Figure 29. Doppler effect of a wave (ko.wikipedia.org).

solid vacuum along the direction of the movement. It moves always to the direction of the more distorted solid vacuum. When the distortion in the advancing direction increases, the velocity increases, namely the object accelerates. This is what we call gravity.

As an object moves at the velocity v, the symmetric distortion of the surrounding solid vacuum changes asymmetrically. If the rest mass is m, then the relativistic mass $m*$ of an object moving at v is given from Eq. (1.4) as

$$m^* = \frac{m}{\sqrt{1 - \frac{v^2}{c^2}}} = \gamma m \; \text{---} \; (2.10).$$

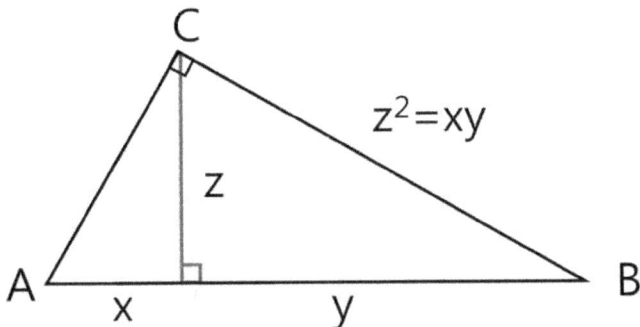

Figure 30. In a right triangle, the product of x and y is z squared.

In Eq. (2.10), γ is the Lorentz factor and a measure of the increased mass due to the distortion of the solid vacuum for an object moving at v.

The hypotenuse of a right triangle is divided into two sections x and y by the vertical straight line from the right corner down to the hypotenuse, as shown in Figure 30. The product of x and y gives the square of the length of the vertical line z. z is then the geometric mean of x and y. If x and y are denoted by $1+v/c$ and $1-v/c$ respectively, γ can be written as

$$\gamma = \frac{1}{\sqrt{1+\dfrac{v}{c}}\sqrt{1-\dfrac{v}{c}}} = \frac{1}{\sqrt{xy}} = \frac{1}{z} \quad \text{---} \quad (2.11).$$

$x + y = 2$ is constant, and z becomes smaller as the difference between x and y increases. In an ellipse

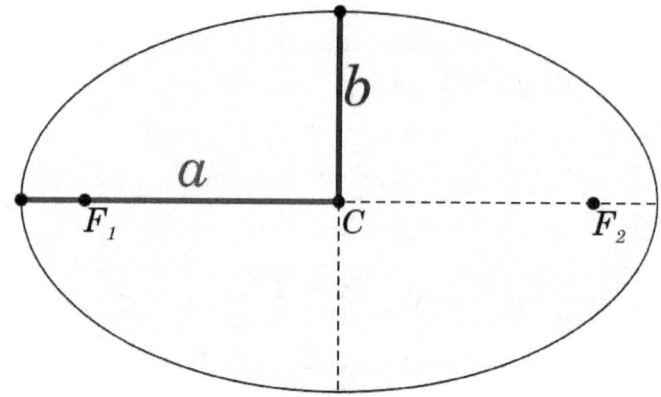

Figure 31. An ellipse whose focal points are F_1 and F_2. The distance from the center C to the ellipse through the focal point is the semi-major axis a and its vertical is the semi-minor axis b.

shown in Figure 31, z means the semi-minor axis b and y is the reciprocal of this value. It is a measure of the distortion of the solid vacuum caused by an object moving at v. If $v = 0$, $x = y = z$ in Figure 30, but as v increases, z decreases and y increases. The kinetic energy of a moving object can be obtained from the energy difference of the relativistic mass and the rest mass of (2.10). Namely,

$$E = (m^* - m)c^2 = (\gamma - 1)mc^2 \quad \text{---} \quad (2.12).$$

For $v \ll c$, $\gamma - 1$ is approximated to $\frac{1}{2}(v/c)^2$ and the kinetic energy becomes $E = \frac{1}{2}mv^2$. The kinetic energy is a quantitative expression of the asymmetric distortion energy stored in the solid vacuum.

Gravity – Point-symmetric distortion of the solid vacuum

Let's go back to the problem of gravity. As shown in Figure 23, when two spherical objects with masses m_1 and m_2 are apart from each other at a certain distance, there should be a difference in the distortion of the solid vacuum at the locations between and outside of the two objects. In the in-between space, the distortion is intensified as the distortion by each object is added, and thus the vacuum lattice spacing will be narrowed. For one object, the distortion of the solid vacuum on the right side differs from that on the left side. As a result, the two objects move toward the stronger distortion of the solid vacuum as with an inertial movement and approach each other. We have an analogous case with the higher speed of sound waves in a solid at higher pressures.[64] As the two objects approach toward each other, the distortion of the solid vacuum between the objects further increases and the probability of energy transfer to this region becomes higher (the individual vacuum lattice oscillates as much as the energy and the probability of energy transfer is higher to the more distorted region) so that the velocity of approach increases. This is the answer to the questions why two physical bodies in the solid vacuum are attracted to each other, and what the nature of gravity is. It is natural that the wavelength of matter wave decreases according to Eq. (2.7) and the velocity increases.

An object moving at a constant velocity v stores energy in the solid vacuum, as shown in Eq. (2.12). (The distortion of the solid vacuum is also the source of mass and of gravity, so that gravitational mass and inertial mass are inevitably the same). This energy is equivalent to the distortion of the solid vacuum on the plane perpendicular to the traveling direction of the object. According to Eq. (2.12), v is expressed as a function of the distortion of the solid vacuum from the relation with kinetic energy, as

$$v = c\sqrt{2\Delta\delta} \quad \text{--- (2.13).}$$

Here we defines $\Delta\delta$ = γ-1 to be the distortion of the solid vacuum. In Figure 29, which shows the Doppler effect, if the distortion of the solid vacuum is elliptical as shown in Figure 31, the distortion of the solid vacuum occurs not only in the direction of movement but also in the normal direction (the perpendicular plane). However, since the distortion in the normal direction is symmetrical about the traveling direction of the object, the distortion is canceled out and does not affect the movement of the object. However, other objects around the moving object will be affected by the distortion of this solid vacuum. In the right triangle of Figure 30, if the hypotenuse is the direction of movement, the distortions of the solid vacuum in the upper and lower directions perpendicular to the hypotenuse will be $\Delta\delta$, but they are canceled out to zero and the distortion of the solid vacuum in the traveling direction of the object

Figure 32. A two-dimensional distortion of the solid vacuum induced by rotation.

will be doubled and becomes $2\Delta\delta$.

However, if an object makes a rotational motion instead of a linear motion, the distortion of the solid vacuum in the normal direction (radial direction) will not cancel each other out. A two-dimensionally symmetric distortion field of the solid vacuum will develop, when an object is rotating as shown in Figure 32. If a particle is near the plane of rotation, it will feel the difference in the distortion in the radial direction, so that a gravitational force is generated by this kind of rotational motion. Therefore, we can understand the orbital movement of all the solar planets concentrated on the plane of rotation of the Sun, as shown in Figure 33.

Now, consider a planet of mass m that orbits at the distance r from the center of the Sun of mass M. When this planet revolves, the gravitational potential energy (PE) due to the Sun and the kinetic energy

Figure 33. All the solar planets orbit around the Sun on its plane of rotation(wikipedia.org). These orbital motions are due to the distortion of the solid vacuum by the solar rotation.

(KE) for the planet to escape at the radial velocity u out of the Sun should be the same. That is,

$$KE = \frac{1}{2}u^2 = \frac{GMm}{r} = PE,$$

to give the escape velocity

$$u = \sqrt{\frac{2GM}{r}} = c\sqrt{\frac{r_i}{r}} \quad \text{---} \quad (2.14).$$

Here r_i is also known as the Schwarzschild radius.[*59] Comparing Eq. (2.13) with Eq. (2.14), the distortion of the solid vacuum in the radial direction decreases with the distance from the center by $2\Delta\delta = r_i/r$. The distortion of the solid vacuum in the circumferential direction perpendicular to the radial direction is $\Delta\delta$, i.e. $r_i/2r$. Since the velocity in the radial direction is $u = dr/dt$, the acceleration α in the radial direction is obtained by differentiating Eq. (2.14) respective to time t as

$$\alpha = \frac{du}{dt} = u\frac{du}{dr} = -\frac{c^2}{2}\frac{r_i}{r^2} = -\frac{GM}{r^2} \quad --- \quad (2.15).$$

It is seen that α has a negative value and is opposite to the direction of u. Multiplying this by mass m of the planet, we have the same equation as shown in Eq. (1.5), Newton's theory of gravity. Here, the gravitational constant G is an index representing a mechanical property of the solid vacuum in our vacuum paradigm.

As the triangle in Figure 30 shows, the larger the deformation (the larger the difference between x and y), the faster the object moves. Deformation of the solid vacuum is developed on the plane perpendicular to the direction in which the object advances. An

*59) The Schwarzschild radius is a parameter in the Schwarzschild solution to General Relativity. It represents the event horizon for Schwarzschild black holes. Borrowed the name of German astrophysicist Karl Schwarzschild. See Section 3.7 for details.

object in rest vibrates around the center of mass if there is no external force. If the spacing of the virtual vacuum lattice on one side is narrower than that of the opposite side due to the distortion of the solid vacuum, there is a higher probability for the object to move toward the narrower lattice spacing, and therefore, a net displacement of matter to one direction will occur as a result of the vibration. This is well described by Eq. (2.8) which describes the behavior of matter wave. The mass and velocity of the object determine the wavelength of matter wave, which is the spacing of the virtual vacuum lattice, i.e. the de Broglie wavelength.

Since the distortion $\Delta\delta$ of the solid vacuum in the direction perpendicular to the radial direction is $r_i/2r$, the velocity will be $\sqrt{2}$ times the radial velocity. That is, the orbit velocity v_O is given by Eq. (2.14) as

$$v_O = c\sqrt{\frac{r_i}{2r}} = \sqrt{\frac{GM}{r}} \quad \text{---} \quad (2.16).$$

If the eccentricity[*60)] is not large, the orbit is close to be a circular one. v_O is then approximated from Kepler's third law as

$$v_O = \frac{2\pi a}{T} = \sqrt{\frac{GM}{a}} \quad \text{---} \quad (2.17)$$

[*60)] The orbital eccentricity is a factor indicating the extent to which the orbit is not a circle. It is 0 for circle, between 0 and 1 for ellipse, 1 for parabola, and more than 1 is for a hyperbolic orbit. It is 0.017 for Earth and 0.206 for Mercury.

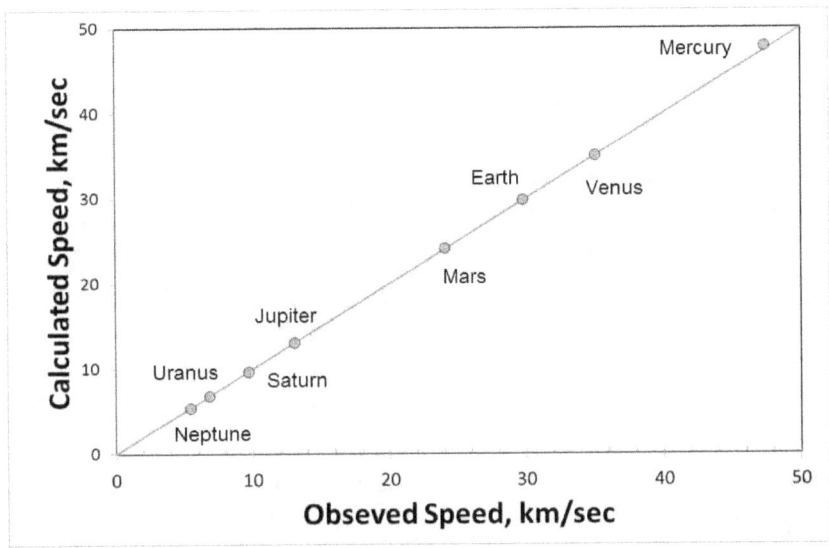

Figure 34. Observed and calculated values of the orbital velocity of the solar planets. The closer to the Sun is the greater the orbital velocity in the order of Mercury, Venus, Earth, Mars, Jupiter, Saturn, Uranus, and Neptune.

with the period T and the semi-major axis a of Figure 31.[65] Eqs. (2.16) and (2.17) are roughly matched. However, Eq. (2.16) is deduced from the inertial motion of the planet, so that gravity and orbital motion can be understood in terms of the inertial motion in the solid vacuum. The above two equations are fully consistent if the orbital radius is the semi-major axis a.

The observed and calculated orbital velocities for the solar planets are compared in Figure 34. The calculated values in the regime of our vacuum paradigm coincide well with the measured ones.

Here we may raise one question. If a planet tries to move away against the Sun's gravitational force at the velocity u, should not the distortion of the solid vacuum in the circumferential direction be left-right balanced so that it does not move? Two answers to this question are plausible. The distortion of the solid vacuum formed in the forward and backward directions of the movement of the planet is canceled by one revolution of the planet around the Sun. Or according to Eq. (2.16), the distortion of the solid vacuum formed in the radial direction by the revolution of the planet precisely cancels the distortion of the solid vacuum formed by the Sun.

2.7. Light deflection in the distorted solid vacuum

As illustrated in Figure 35, it is well explained by General Relativity that light deflects when passing by massive objects such as the Sun. As shown in Figure 12, the deflection of light by the Sun was proved by Eddington's observations in the total solar eclipse on May 29, 1919,[66] and was further confirmed to be 1.75±0.09" by additional observations, as exact as predicted by Einstein.[67] Here we also calculate the deflection of light based on our vacuum paradigm.

Derivation of the light deflection in the new paradigm

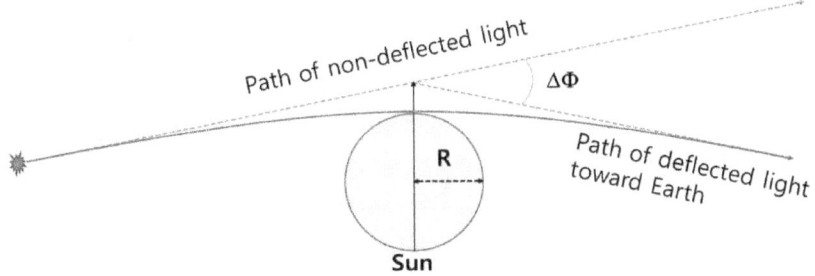

Figure 35. Deflection of light by the Sun.

When light from a distant star travels toward the Sun, the propagation of light through the solid vacuum can be broken down into the two components, one for the radial direction and one for the circumferential direction of the Sun, as shown in Figure 36. The distortion of the solid vacuum in the radial direction due to the influence of the solar mass is compressive one $2\Delta\delta = r_i/r$. Therefore, while moving forward by Δp (AC) for a time interval Δt as shown in Figure 36, it approaches the center of the Sun by Δr (AB) and moves in the circumferential direction of the Sun by Δq (BC). If there is no distortion of the solid vacuum, it should be $\Delta q + \Delta s$ (BD) instead. Δs (CD) is the relative slowing of the light propagation due to the distortion of the solid vacuum in Δt, and this value is given as

$$\Delta s = \Delta q \times 2\Delta\delta = \Delta q \times \left(\frac{r_i}{r}\right) \quad --- \quad (2.18).$$

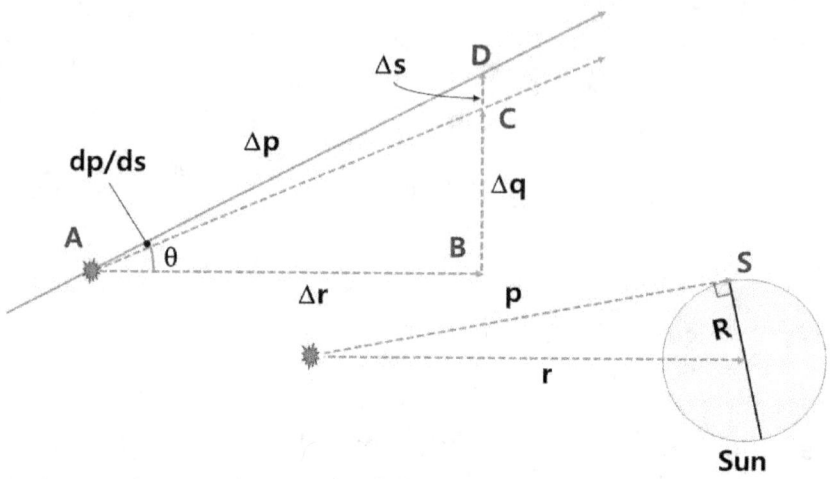

Figure. 36. When light travels toward the Sun, the propagation vector can be divided into the component Δr in the radial direction (to the center) of the Sun and the component Δq in the circumferential direction. The speed of light in the radial and circumferential directions will be different due to the difference in the distortion of the solid vacuum influenced by the solar mass. It is Δs.

The distances r and p have the following relationship,

$$r^2 = p^2 + R^2 \rightarrow p\Delta p = r\Delta r \quad \text{--- (2.19)},$$

where R is the solar radius. From Eq. (2.19) and Eq. (2.18), and the relation $\Delta p^2 = \Delta r^2 + \Delta q^2$ from the geometry of Figure 36, we have

$$\Delta s^2 = \Delta p^2 \left(\frac{r^2 - p^2}{r^2}\right)\left(\frac{r_i}{r}\right)^2 = \Delta p^2 \frac{R^2}{r^2}\left(\frac{r_i}{r}\right)^2$$

so that

$$\Delta s = \frac{r_i R}{r^2}\Delta p = \frac{r_i R}{p^2 + R^2}\Delta p \;\; \text{---} \;\; (2.20),$$

or

$$s = \int_{-\infty}^{0} \frac{r_i R}{p^2 + R^2}dp = \frac{\pi}{2}r_i \;\; \text{---} \;\; (2.21).$$

For Eq. (2.21) we used $\int \frac{1}{x^2 + a^2} = \frac{1}{a}tan^{-1}\frac{x}{a}$. The angle when light is closest to the Sun is when $p = 0$ or when $r = R$. The deflection angle in radian is ds/dp and the deflection angle φ at the solar surface ($p = 0$) is given from Eq. (2.20) as

$$\phi = \frac{ds}{dp} = \frac{r_i}{R} \;\; \text{---} \;\; (2.22).$$

This deflection angle will be twice for an Earth observer, as shown in Figure 35, since light should experience the same effect as it reaches the Earth observer. The total deflection is then

$$\Phi = 2\phi = \frac{2r_i}{R} = \frac{4GM}{c^2 R} \;\; \text{---} \;\; (2.23).$$

This result is the same as that derived from General Relativity.[68] In the theory of relativity, space and time

- 99 -

in the gravitational field are distorted and light is deflected. Light is a concept in which its speed does not change and time and space are dilated and contracted. In the regime of our vacuum paradigm, there is no concept of spacetime distortion, and only the speed of light due to distortion (compression or relaxation) of the solid vacuum changes. There is no strange concept that time is getting longer and shorter. If the behavior of light is the same as the behavior of sound waves, from the fact that the speed of sound waves is higher at higher pressures, light will speed up where the solid vacuum is compressed. Eq. (2.23) proves this hypothesis.

2.8. Precession of Mercury

When two objects with different masses revolve relative to each other, the small object rotates in an elliptical orbit around the large one. Because it rotates around the center of gravity of the two objects. In Newton's mechanics, the perihelion[61] does not move when only considering the relative motion of the two objects. In the solar system, however, the perihelion of Mercury to the Sun is not fixed but rotating for various reasons. This is the precession of the planet and the main cause is the influence of the nearby other planets.[62]

---

[61] the periapsis or it is the perihelion in the solar system.
[62] Some are due to the flattening of the Sun, but it is negligible.

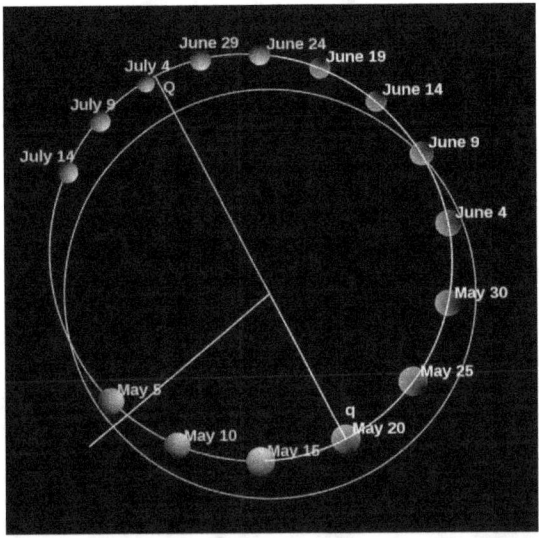

Figure 37. Mercury's orbit: The point where the two lines intersect is the center of the Sun (wikipedia.org).

Mercury's orbit is deviated from the center of the Sun, as shown in Figure 37, and the perihelion rotates by 574" for 100 years. Figure 38 shows an exaggerated perihelion movement of Mercury. In the regime of Newtonian mechanics the precession of Mercury is calculated as 532" by taking the gravitational effects of the other planets into account. However, there is still a difference of 43" from the observed value.[69] This unusual precession of Mercury was first noticed in 1859 as a problem of astrodynamics by Le Verrier.[*63] He analyzed

*63) Urbain Le Verrier (3.1811 - 9.1877) French mathematician,

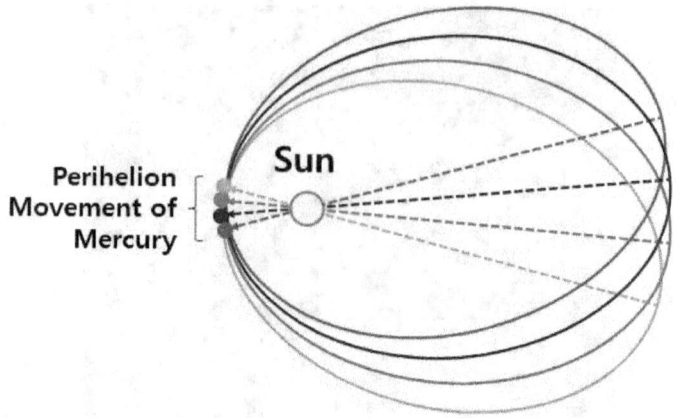

Figure 38. The precession of Mercury.

astronomical data from 1697 to 1848 and found that
the precession had a difference of 38"/100 years*64)
from the Newtonian calculation. Thereafter several
hypotheses appeared and attempted to interpret this
difference, but all but General Relativity failed.

Precession of Mercury by General Relativity

In General Relativity, Mercury's directional change in
the elliptical orbital axis is explained by the change in
the spacetime curvature due to gravity. According to
the theory, while Mercury is passing through the

astronomer. He is best known for predicting the existence and
position of Neptune using only mathematics. Neptune was found
by J.G. Galle (6.1812 - 10.1910, German astronomer) in 1846.
*64) It was corrected to 43" later by S. Newcomb (3.1838 -
7.1909), Canadian-American astronomer.

Table 3. Origin of the precession of Mercury

Origin of the precession	precession ("/100 years)
Gravitational tugs of other bodies	532.3035
Flattening of the Sun(quadrupole moment)	0.0286
Distortion of spacetime due to the general relativity	42.9799
Others	-0.0020
Sum of calculation	575.31
Observed value	574.10±0.65

gravitational field of the Sun, time goes slow and length is contracted.

Mercury moves around the distorted spacetime around the Sun, so that Mercury's orbit should be curved and the perihelion moves accordingly. The difference between the measured precession and the predictions of Newtonian mechanics can be accessed by calculating the effect of the distortion of spacetime. The current precession of Mercury obtained using radar technology is 574.10±0.65"/100 years,[70] and the causes and predictions for the precession are shown in Table 3. The calculated value considering the spacetime distortion by General Relativity makes well up the difference.[71] Other planets also do precession, though weaker than Mercury does. It is 3.84"/100 years for Earth, 8.62"/100 years for

Venus,[72] and the precession of the recently discovered double pulsar *[65] system PSR 1913+16*[66] is as high as 4.2°/year. Their precession values are in good agreement with the predicted by General Relativity, thus enhancing the reliability of this theory.[73]

Einstein calculated the precession ε based on General Relativity for an elliptical orbit of a planet whose semi-major axis is a, eccentricity is e, and periodicity is T as follows.[74]

$$\epsilon = 24\pi^3 \frac{a^2}{T^2 c^2 (1 - e^2)} \quad \text{---} \quad (2.24),$$

where c is the speed of light. For Mercury, $a \cong$ 5.79×10^{10} m, e = 0.206, and T = 87.97 days. Inserting these values into Eq. (2.24), the precession for one period is 5.028×10^{-7} radians (2.88×10^{-5} degrees or 0.104"). Mercury is revolving about 415 times per 100 years, so that the relativistic precession is calculated as about 43" for 100 years.

Precession, the process of releasing the distortion

In our new vacuum paradigm, kinetic energy and gravitational potential energy are substantially the same distortion energy stored in the solid vacuum. In

---

*65) A pair of pulsating radio stars. They are neutron stars with very high magnetic field densities and rotational speeds. The pair emits pulsating electromagnetic waves in a period of 0.001 to 10 seconds.
*66) It is also called the Hulse-Taylor binary.

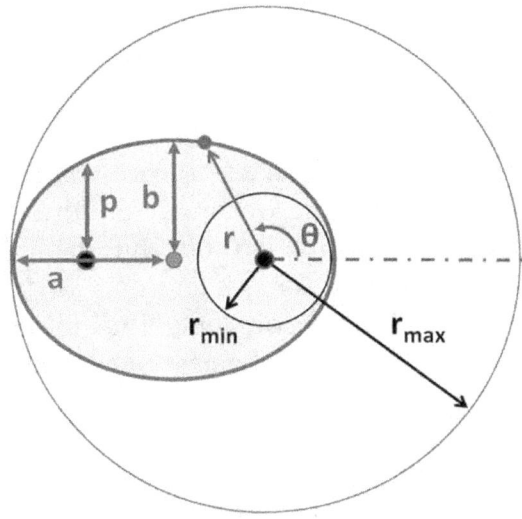

Fiture 39. Coordinate system showing
an elliptical orbit (wikipedia.org).

this context, it is very simple and clear to understand
the perihelion movement of the binary system
including the Sun and Mercury in terms of the
distortion of the solid vacuum. When Mercury rotates
one revolution around the Sun, the associated
distortion is the radial one of r_i and the
circumferential one of $\frac{1}{2} r_i$, as evident from Eq. (2.14)
and Eq. (2.16). If the orbital radius is r, and if there
is no distortion of the solid vacuum by the Sun, the
radius should be $r - r_i$, to give a difference in the
orbital length of $2\pi r_i$. The orbital motion of Mercury
also makes a contribution to the orbital length of πr_i.
As a result, a total of $3\pi r_i$ due to the distortion of the

solid vacuum is included in one orbital revolution of Mercury. To release this distortion, the axis connecting the center of the Sun and Mercury should rotate by the length of $3\pi r_i$ for each orbital rotation. This is the reason of the precession of Mercury in the Sun's gravitational field. In order to obtain this value in radians, it must be divided by the orbital radius. Namely, the precession is given as $3\pi r_i/r$. Since Mercury has an elliptic orbit and the precession is observed on the perihelion basis, the radius of the orbit should be the harmonic mean[*67] r_h of the perihelion r_{mn} and the aphelion r_{mx} of the elliptical orbit. In the elliptic orbital coordinate system of Figure 39, p is called the semi-latus rectum and is equal to the harmonic mean in the following relation.

$$\frac{1}{r_{mn}} - \frac{1}{p} = \frac{1}{p} + \frac{1}{r_{mx}} \rightarrow p = \frac{2r_{mn}r_{mx}}{r_{mn} + r_{mx}} = r_h \ \text{---} \ (2.25).$$

Inserting the following equation, Kepler's third law,

$$\frac{T^2}{a^3} = \frac{4\pi^2}{G(M+m)} \ \text{---} \ (2.26),$$

into Eq. (2.24), we have

[*67] The harmonic mean is the reciprocal of the arithmetic mean of the inverse of given numbers. It is mainly used to determine the average rate of change.

$$\epsilon = \frac{6\pi G(M+m)}{ac^2(1-e^2)} = \frac{2G(M+m)}{c^2} \frac{3\pi}{a(1-e^2)}$$

or

$$\epsilon = \frac{3\pi r_i}{a(1-e^2)} = \frac{3\pi r_i}{2\left(\dfrac{r_{mx} r_{mn}}{r_{mx}+r_{mn}}\right)} = \frac{3\pi r_i}{r_h} \quad \text{---} \quad (2.27).$$

Eq. (2.27) is the precession of Mercury predicted in the regime of our vacuum paradigm. This is exactly the same as the value obtained based on General Relativity. Calculating Mercury's precession is very complicated.[75] However, we could determine the precession very simply in terms of the distortion of the solid vacuum. Newton's theory of gravity and General Relativity were basically obtained in milieu of the "empty" space of the universe, and in these theories gravity is a fundamental force that exists by itself in nature. The vacuum in our vacuum paradigm is not a fluid like "Ether" but made of a highly dense ordered lattice structure and the relative displacement of this lattice is the origin of energy existing in the form of vibration (in the form of electromagnetic waves and matter wave or neutrinos).

Gravity is an inevitable phenomenon caused by the distortion of the solid vacuum due to the presence of matter (such as the Sun or Earth) and the characteristics of matter wave. Gravity is not a groundless thing. Gravity does not exist by itself. The spacetime distortion in the presence of mass, as mentioned in General Relativity, is only a different wording of the distortion of the solid vacuum. General

Figure 40. When a meteor moves very fast, they are distorted and move discontinuously due to the resistance of the solid vacuum. The largest meteor observed in Western China on August 13, 1978. It is a typical Perseus meteor shower (from a Korean science journal, 과학동아, Jan. 1993).

Relativity and the approach in our vacuum paradigm are fundamentally different, but the results are the same in some way. In General Relativity, spacetime is distorted because there is mass. However, in our vacuum paradigm, energy input into the solid vacuum makes the vacuum lattice vibrating, resulting in an attractive force, the "so-called" gravitational force. Since the vacuum is made of a dense medium, it is highly resistant to severe distortion. Eq. (2.9), the energy stored in the solid vacuum increases exponentially as the velocity of an object approaches the speed of light. The resistance of the solid vacuum to deform can be confirmed from the behavior of meteors as shown in Figure 40. When a meteor moves very fast, the meteor is distorted and appears to be discontinuous due to the resistance of the solid vacuum. This makes a fundamental difference from General Relativity, which assumes that there is nothing in the vacuum, yielding contradictions to distant cosmic phenomena, such as dark energy, dark matter, and black holes. These are discussed in the next chapter.

III. New thinking on the universe

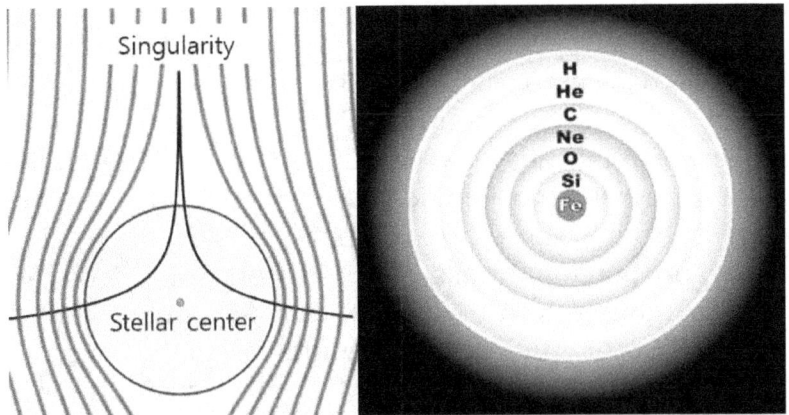

The internal stress of a spherical object placed in the solid vacuum increases exponentially toward the center and can form a singularity as shown in the left figure. Because of high pressures at the stellar center, nuclear fusion occurs actively and heavy elements such as iron are synthesized. The stellar interiors just before the explosion to a supernova have a onion-like elemental distribution as shown in the right figure, and heavier elements are concentrated in the central region.

3.1. Where comes the energy of the universe?

As discussed in Chapter 2, the movement of matter distorts the cold solid vacuum and induce additional movement of nearby objects due to this distortion. Inertial motion of an object accompanies distortions of the solid vacuum not only in the forward direction but in the backward direction, so that kinetic energy is stored in the solid vacuum. When an object rotates, there is a difference in the distortion of the solid vacuum between the outside and the inside of the rotational orbit. Therefore, objects outside the orbit will be attracted toward the region of stronger distortion, namely toward the center of rotation. If the lattice points of the solid vacuum oscillate without directionality, the distortion of the solid vacuum will develop in a three-dimensional point symmetric mode. This is the nature of mass and it is nothing but the vibrational energy of the solid vacuum. Due to this point symmetric oscillation the distortion of the solid vacuum develops point-symmetrically. The strength of the distortion is inversely proportional to the distance from the center as shown in Eq. (2.14), and the induced acceleration is inversely proportional to the square of the distance. It is exactly Newton's theory of gravity. In the regime of our vacuum paradigm, mass is nothing but the vibration energy, and when the symmetry of this vibration is broken, the central point of vibration moves, through which the frequency of matter wave increases.

Energy of the universe is trembling of the solid vacuum

The Sun is a lump of energy, as with most of other stars. When the amount of energy is large, its mass is also large and the solid vacuum around it is distorted as much as the amount of energy. This distortion induces the corresponding stress in the stellar interiors. Figure 22 shows the stress distribution formed inside a spherical body. Its center is in a very high pressure state. In elastic theory, there is a singularity in which the stress at the central point becomes infinite.

The strong force, electromagnetic force, weak force, and gravitational force, the four kinds forces in nature can be estimated in terms of the relative strength to the electromagnetic force as 20 for the strong force, 10^{-7} for the weak force, and 10^{-38} for the gravitational force. Assuming that the difference in the strength is due to the stress difference along the distance from the center as shown in Figure 22, the range (distance) of these forces can be calculated relatively to the strong force (when the force is inversely proportional to the square of the distance). The strong force is the force acting among protons and neutrons in atom nuclei. If the diameter of the strong force is in the range of 10^{-15} m, the electromagnetic force is 1/20 of the strong force, so the diameter of the force range is 4.47×10^{-15} m, 1.41×10^{-11} m for the weak force, and 4.47×10^{4} m (~45 km) for the gravitational force, respectively. If the electromagnetic force is acting on the electrons and

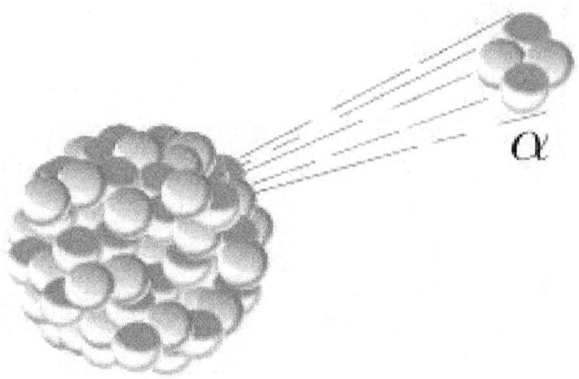

Figure 41. Alpha decay, a type of radioactive decay (wikipedia.org).

protons in atoms, the force range will be about 5 times the range of the strong force. The range of the weak force is similar to the radius of one atom (helium, covalent radium, 2.8×10^{-11}m).[76] Therefore, this force is due to the stress around the boundary of an atom. The weak force is associated with radioactive decay, which reduce nuclear energy during nuclear fission.[*68] The weak force is a kind of stress around an atom whose energy is released via radioactive decay, as shown in Fig. 41.

As with the strong force, the center of a spherical star will be subjected to very high stresses and the behavior of atoms constituting the star in this very

*68) Figure 41 illustrates α decay, a type of radioactive decay. When an alpha particle is released from an atomic nucleus, the mass number of the atom is reduced by 4 and the atomic number is decreased by 2.

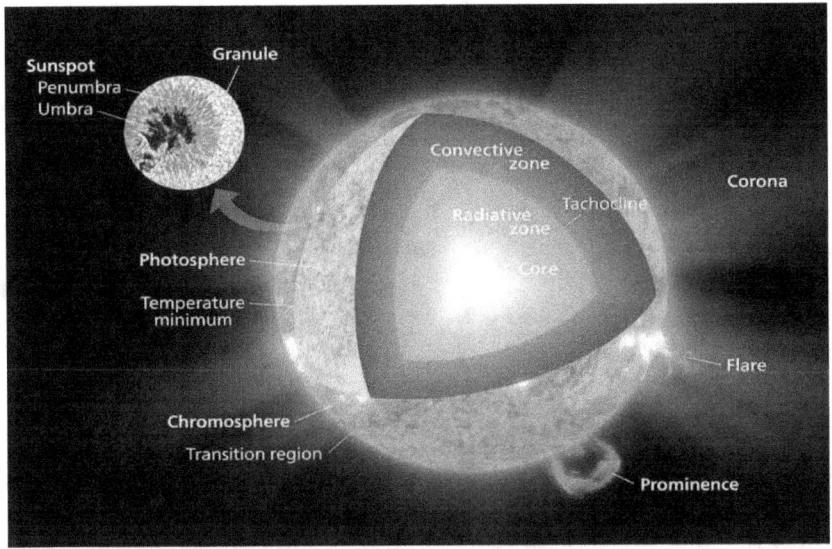

Figure 42. The inner structure of the Sun (wikipedia.org).

high stress field is expected to be very unusual. In this regard, we review the structure of the Sun and some solar planets, which are expected to have very high stress in the core, and investigate whether the rotation and orbits are related to this stress. We will also review the distant cosmological phenomena in the regime of our vacuum paradigm.

3.2. The Sun in the solid vacuum

The Sun is the nearest star to us. Light can reach us just in 8 minutes. Earth revolves around the Sun and

Earth's life is sustained by the energy emitted from the Sun. But how does the Sun keep emitting enormous amounts of energy permanently? Of course, all stars have lifetime, their energy is finite, and eventually the energy of the Sun will be exhausted, and become a star like a white dwarf[*69] in the long run. As we will discuss on the lifetime of stars sometime later, here we will look into the structure of the Sun, the rotation, orbit, and internal convection in the regime of our vacuum paradigm.

Structure of the Sun

The Sun is composed of 74.9% of hydrogen and 23.8% of helium by weight, and contains trace elements such as iron.[77] The structure of the Sun is shown in Figure 42. The core reaches about 20-25% of the solar radius from the solar center.[78] The density is about 150 times that of water and the temperature is 15.7 million K.[79] According to a recent analysis, the core is rotating faster than the upper radiative zone.[80] The Sun produces energy by fusion in the core, in which hydrogen turns into helium, as shown in Figure 43. In the radiative zone, reaching the radius of 70% from the solar center, thermal radiation occurs that transfers the energy of the core to the outside. There

---

*69) An old star exhausted in energy via nuclear fusion. Stars with sub-medium mass become red giants, even if the core collapses and rises in temperature and pressure, not reaching enough temperatures to induce carbon fusion, then the outer atmosphere is released into outer space to form nebulae, only the core and oxygen remain to form white dwarfs.

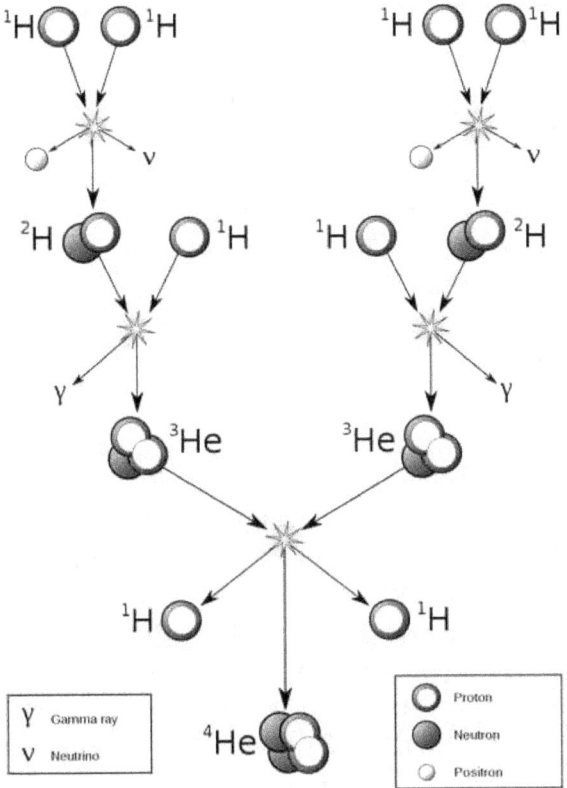

Figure 43. The p-p chain reaction (wikipedia.org): Helium is generated by hydrogen fusion in the Sun.

is no heat convection, and internal matter is cooled as it goes outward. Temperatures range from 7 to 2 million K. The density drops from 20 g/cm^3 to 0.2 g/cm^3, going from the bottom of the radiative zone to the top surface.[81] In the outer layer of the Sun, to the depth of 200,000 km from the solar surface, the solar plasma is low in density and temperature, and does not transmit its internal thermal energy out via radiation. As a result, a convective zone is forming in which the ascending stream moves the hot material up to the photosphere of the Sun. The upward flow in the convection layer forms granules on the solar surface.

What happens inside the Sun – squeeze hydrogen, generate helium

Summarizing the structure and movement of the Sun, the core rotates at a very high speed and produces energy by fusion, while the Sun's radiative zone carries the energy of the core and rotates itself, but its rotational speed is slower than that of the core. The convection layer conveys heat to the exterior via convection (of plasma). In the regime of our vacuum paradigm, the Sun should be in a very high stress state at the center, as similarly shown in Figure 22, due to the influence of the surrounding solid vacuum. The distortion will become rapidly severer as the distance from the center decreases as discussed in Section 2.6. Due to this severe distortion, hydrogen atoms in this vicinity cannot withstand high stresses, electrons and protons will turn into neutrons, and stable helium atoms are produced in this process. This is known as the proton-proton (p-p) chain reaction as shown in Figure 43 and is known as the fusion reaction that occurs in the solar core.[82] The fusion energy is released and thus the core temperature is very high.

If four hydrogen atoms are involved in the formation of a helium atom, two electron-proton pairs are consumed and two neutrons are produced. In this process, the potential energy between electrons and protons will be emitted as light and neutrinos,[*70] which is the very nature of the solar energy. Most neutrinos arriving Earth are those produced by the nuclear reaction in the Sun. 65 billion (6.5×10^{10}) neutrinos per square centimeter reach Earth every second. Neutrons also carry some energy, squeezed out in the environment of ultra-high temperatures and pressures of the solar center. As mentioned in section 2.4, atoms with the higher atomic weight are composed of neutrons with lower energy, and thus their number of surplus neutrons are higher as shown in Figure 25. When a star explodes into a supernova while generating heavy elements inside the star, neutrons with high energy will be exposed directly to the solid vacuum and radioactively decay into protons and electrons (+ antineutrinos).[83]

The Sun is still alive, and helium produced in the

---

*70) We will define that the source of matter is electromagnetic waves and neutrinos, and when the energy of these waves increases, electrons and positrons are generated, and positrons are captured at the vacuum lattice points to be protons. If electrons and protons react, they combine to release energy in the form of electromagnetic waves and neutrinos, and some of the remaining energy is trapped in the vacuum lattice as neutrons. Neutrons which have still high energy are those neutrons observed in nuclear fission.

fusion process will turn into heavier elements under ultrahigh temperatures and pressures in the solar core. It cannot be ruled out that the core is supposedly in a highly dense (high-temperature) fluid form due to this nuclear reaction. Or if all the energy is exhausted, the very central core may be composed of matter similar to the cold solid vacuum.

Rotation and revolution of the Sun through the solid vacuum

The solar core is rotating faster than its outer radiative zone. The rotational speed is 7.189×10^3 km/h (about 2 km/s) at the equator.[84] The central core rotates about 4 times faster than that,[85] but the average line speed is similar because the distance from the center is 3/4 shorter than that of the surface. This rotation further distorts the solid vacuum on the plane perpendicular to the axis of rotation and causes an asymmetry in mass and gravity. The Sun is an oval whose flattening is 9×10^{-6} and whose equatorial diameter is 10 km longer than the pole one. The reason why the Sun is not a perfect sphere is presumed to be the internal rotation with directionality, but it can be influenced by the disk-shaped structure of our galaxy as shown in Figure 44 or by the solar planets orbiting in the plane of the Sun's rotation.

The stellar core is very pressurized and nuclear fusion occurs very actively and the energy generated by fusion is emitted. The ejected energy can propel

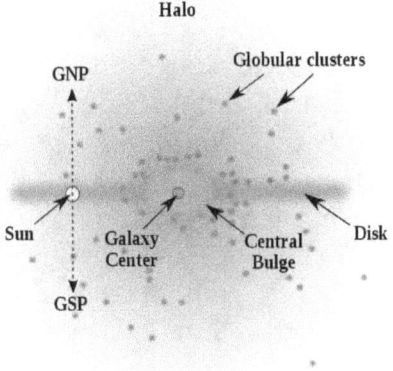

Figure 44. The shape of our galaxy and
the location of the Sun (wikipedia.org).

the inner core of the Sun, as a balloon gets a propulsion force when the internal air is discharged. The inner core will move with the propulsion force, and if its vector sum is not 0, a net torque will be developed due to the unbalance of the amount of released energy. This is thought to be the main reason for the rotation of the Sun (and other stars, too). Since the Sun is not a solid, the rotational speed of the core and the convection rate of the Sun's radiative zone should be different, and accordingly the rotational speed of the Sun's radiative zone is lower than that of the solar core. Since the Sun is a fluid composed of hydrogen and helium, how does the centrifugal force of rotation cause the Sun to deform like a disc? The Sun revolves around our galaxy with an average speed of 8.28×10^5 km/h (230 km/s).[86] The solar orbital motion will inevitably leads to another distortion of the surrounding solid vacuum. Namely, a compressive stress is additionally generated in the forward direction of the orbit to suppress the deformation due to the rotation of the Sun. The rotation and the orbital motion of the Sun seem to work to maintain the current flattening.

Meanwhile, just as the Sun is revolving around our galaxy, the solar planets revolve at the velocities that are proportional to the square root of the solar mass, as shown in Eq. (2.17). Applying the same criterion to the Sun, the mass of our galaxy is estimated to be 9.99×10^{10} times the solar mass, $M_\odot$, based on Eq. (2.17) with the distance to the galactic center, 26,500 light years.[87] The mass inside the orbit is only counted for the orbital motion. So from the orbital motion of the stars in the outskirts of the galaxy the total mass of our galaxy is estimated to be 7×10^{11} M$_\odot$.[88] The Sun is not in the central region, nor in the outskirts of the galaxy, as shown in Fig. 44. Thus, the mass of the galaxy outside the Sun's orbit is six times larger than that inside the Sun's orbit, since 7×10^{11} M$_\odot$ - 9.99×10^{10} M$_\odot$ = 6×10^{11} M$_\odot$. However, referring to Figure 43, considerable masses appear to be concentrated around the galactic center, and the mass present outside the Sun's orbit appears to be relatively small. It is known that there is a very massive black hole at the galactic center.[89] Thus, the more mass outside the Sun's orbit suggests the presence of invisible "dark matter". In the regime of our vacuum paradigm, black holes are not very massive things, but just empty spaces absolutely free of energy or of mass.[*71] In this case, the high mass in the outer region can be understood without relying on dark matter.[*72] However, it is also conceivable that the mass of the galaxy's outer shell, calculated by Eq. (2.17), can be high because of the additional mass due to the distortion of the solid vacuum by the rotation of the stars in the outer shell of the galaxy.

3.3. A cold star - Earth

The solar planets are orbiting the Sun not exactly but

[*71] See Section 3.7 for black holes.
[*72] Dark matter is discussed in detail in Section 4.2.

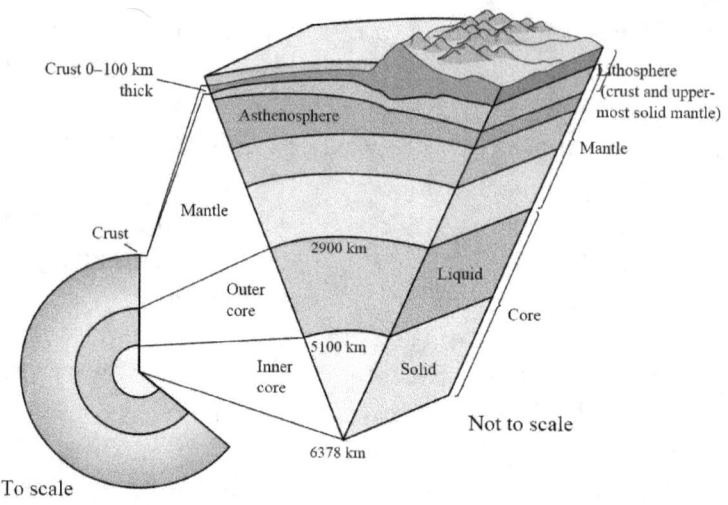

Figure 45. Inner structure of Earth (wikipedia.org).

approximately on a plane called the ecliptic plane. Of the eight planets, Earth is closer to the Sun after Mercury and Venus. The average orbital speed is 1.072×10^5 km/h (27.8 km/s), with the orbit length around 924,375,700 km revolving around the Sun for 365.256 days.[90] Earth is not a true sphere, like the Sun. It has the equatorial radius of 6378.1 km and polar radius of 6356.8 km. The equatorial radius is 21.3 km longer than the polar one. The difference is large, compared with the Sun of 5 km. The rotational period of Earth is 23.934 hours (not exactly 24 hours). The rotational speed at the equator is 1674 km/h (465 m/s) and the axial tilt is 23.439 281°. Earth is composed of materials much denser than the Sun and has an average density of 5.5 g/cm^3. More than 90% of the entire Earth is composed of a mantle and a core. The detailed structure is shown in Figure 45.[91]

Inner structure of Earth

The internal structure of Earth is layered. The outermost part is composed of a crust, followed by a mantle and a core. The crust is very thin in thickness compared with Earth's radius, and its structure and composition vary greatly depending on the region. The mantle begins at the bottom of the crust, the Mohorovici*[73] discontinuity, reaching the depth of 2,900 km. The mantle is made of hard rocks, but it acts like a very slow moving fluid, causing convection. Unlike the mantle or crust, which consists of minerals, the core is made of iron and nickel. The core is divided into two layers, the outer core and the inner core at the depth of 5,100 km. The outer core is in a liquid state (density 9.9~12.2 g/cm^3) because S-waves*[74] cannot propagate, and the inner core is assumed to be solid (density 12.8~13.1 g/cm^3).[92] The radius of the inner core is 1,220 km, and the outer core is a liquid of iron and nickel surrounding the

---

*73) Andrija Mohorovicici (1.1857 - 12.1936) Yugoslavian physicist from the Croatian region. He found the Mohorovicic discontinuity.

*74) Secondary waves. It is one of seismic waves and a transverse wave. The speed of a S-wave is 3 to 4 m/s and propagates only in solids. The earthquake damage by S-waves is larger than P waves (primary waves, velocity 7 ~ 8 m/s) due to the large amplitude.

inner core up to 3,400 km from the center. The temperature is estimated to be more than 5,500 K for the inner core and 3,000–5,500 K for the outer core, respectively. The outer liquid core is convective under the influence of Earth's orbital movement and thermodynamics, and it is believed that the strong magnetic field of Earth is maintained by the movement of this conductive fluid. The outer core rotates with a speed similar to Earth's rotational speed, but in a recent research it turns out to be 0.1 to 0.5° faster per year.[93]

If the Sun is an active volcano, Earth is rather an extinct one. It is made up of heavy elements, exhausted ones in energy. Unlike the Sun, iron, nickel and other very heavy elements constitute the core. Referring to section 3.5, heavy elements are synthesized by nuclear fusion under ultra-high pressures inside of giant stars. When these stars explode and turn into supernovae, heavy elements are released and their debris aggregates to be smaller stars. The heavy elements of Earth should have come from supernovae. In Earth, the inner pressure exerted by the surrounding solid vacuum is much lower than in the Sun (mass is only 3/million of the Sun), and the rate of nuclear fusion in the interiors of Earth will be very low, when considering the temperature of the Sun of 13.6 million K. The solid state of the inner core means that the energy is exhausted, and it can be imagined that matter in the very inner core is similar to the cold solid vacuum itself free of energy. Referring to Figure 25 in Section 2.4, the heavier the element, the greater the number of neutrons than that of protons in the nucleus. In the regime of our vacuum paradigm, surplus neutrons are neutrons which have much lower energy than paired neutrons with protons in atoms or than free neutrons. It is reasonable to assume that the inner core are composed of elements having more surplus neutrons and much higher atomic weight than those found or synthesized at present. It is not simply understood that the inner core with higher temperatures than the outer core is in a solid state. But if the inner core is solid, it would consist mostly of neutrons almost free of energy. The outer core also has high temperatures but is deviated from the very center, so the movement is more fluidly than that of the inner core. This means that energy ejected by nuclear fusion still remains as kinetic energy.

Rotation and flattening determine the fate of Earth

Earth has much a larger flattening[*75)] of 0.0034 than the Sun has.[94] Energy dissipation from fission or fusion in the Sun makes the inner core rotate, which will influence the Sun's flattening. However, because Earth lacks hydrogen, there will be almost no energy supply by nuclear fusion, and since the density is higher than the Sun and composed of heavy elements, the flattening may be smaller than the Sun. Why is it so larger than the Sun? Venus, which is similar to Earth in size, density, and mass has zero flattening.[95]

[*75)] flattening = (equator radius - polar radius)/equatorial radius.

As was previously mentioned about the rotation and flattening of the Sun, we can find the reason for the difference in the rotational speeds of Earth and Venus. The rotational speed of Earth's equator is 1674.4 km/s, while it is 6.52 km/s for Venus.[96] Venus has no internal energy to rotate by itself. Thus, the rotational period of Venus is -243 days, almost the same as the orbital period of 225 days. The energy of Earth's rotation was originated from the fusion energy, and the energy that is not released remains as the rotational energy of heavy elements. This is also the case when neutron stars, the remnants of supernovae, rotate at very high speeds as will be discussed in Section 3.6. As previously mentioned for the Sun, Earth's rotation yields an asymmetrical distortion of the solid vacuum in the plane perpendicular to the axis of rotation and affects Earth's flattening. Due to Earth's rotation and flattening, the gravitational acceleration at the equator is 9.780 m/s^2 and at the pole is 9.832 m/s^2, yielding a 0.5% difference.[97] In terms of our vacuum paradigm, since the surface of the equator is farther from the center due to flattening, the distortion of the solid vacuum is relatively weak and the rotation further reduces the distortion of the solid vacuum near the equator surface. This makes the difference.

Can we consistently explain the relationship between the flattening and rotational periods of the planets other than Venus? Figures 46 and 47 show the flattening and rotational periods of the solar planets, respectively. The rotational periods of Mercury and

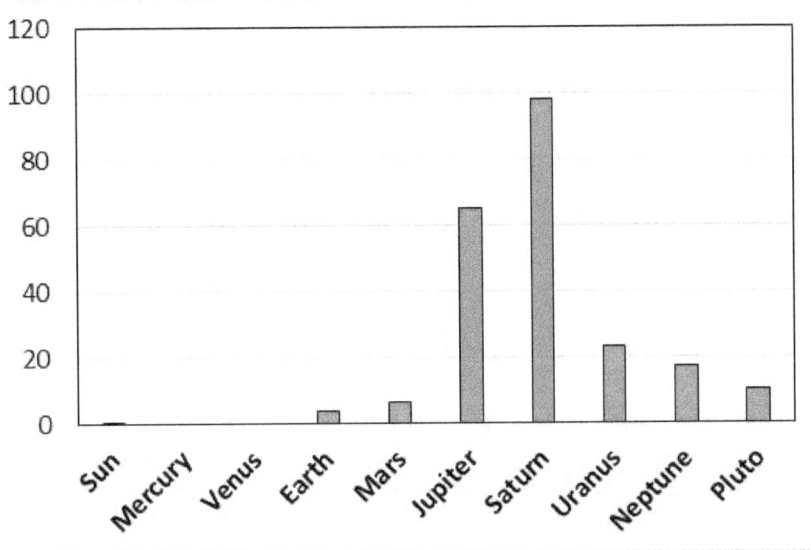

Figure 46. Flattening of the solar planets (unit: 10^{-3}).

Venus with zero flattening is much longer than other planets with higher flattening. In general, the greater the flattening, the shorter the rotational period. The Sun's flattening is very small except Mercury and Venus, while the rotational period is relatively long. Since the solar density is lower than that of Earth, the rotational force (torque) of the core is difficult to be transmitted to the entire star, so the rotational period is different from that of the solar core. The low flattening of the planets suggests that their rotational speeds of the cores are also low.

In Earth the rotational speed of the outer core is higher than that of the mantle, which will inevitably

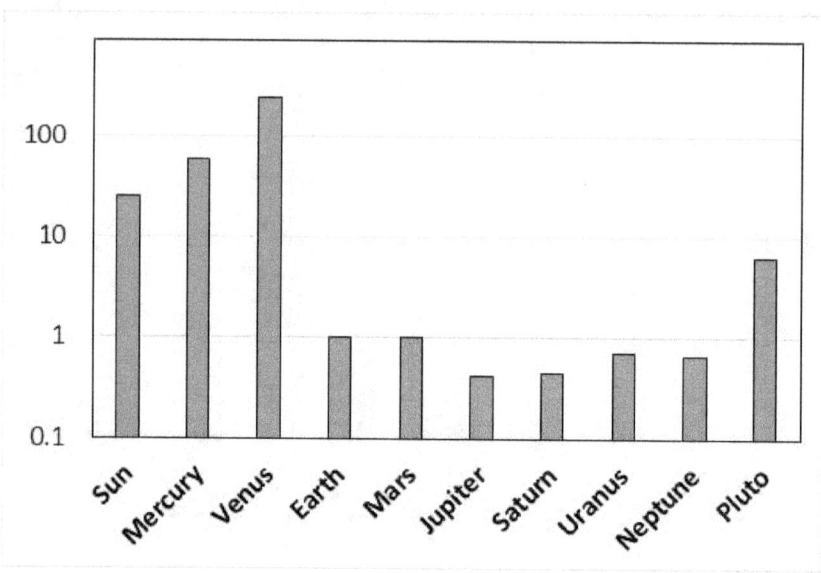

Figure 47. Rotational periods of the solar planets (unit: days).

generate frictional heat at the interface. The heat will be released out of Earth. Earthquakes and volcanic activities are examples of such a energy release. Earth thus consumes its internal energy. Earth will gradually lose the energy of rotation, eventually yielding zero flattening. On December 4, 2006, the Hankyoreh Internet Newspaper (a Korean Newspaper) wrote that Earth's rotational speed would gradually decrease, Earth rotates 25 hours a day after 380 million years and stops its rotation completely after 7.5 billion years.[98] The media also wrote, contrary to our estimate, that the main cause of the slowing down of

Earth's rotation is the 'tidal force' caused by the Moon. It is not clear yet, which argument is true, but it is clear that Earth is getting slower with its rotation eventually to a completely stop.

3.4. Jupiter resembling the Sun

Jupiter is the largest solar planet. As shown in Figure 46 and Figure 47, Jupiter's flattening is about 20 times that of Earth and the rotational period is shorter than that of Earth (about 0.4 days or 9.9 hours), suggesting that Jupiter has still a lot of energy and is much younger than Earth. It is similar to Saturn in this regard. Its mass is about 1/1000 times the solar mass and, as with the Sun, it consists mainly of hydrogen, and helium accounts for 25% of the total mass.[99] Jupiter's mass is 2.5 times the total mass of all other solar planets, but its density is as low as 1.326 g/cm^3 (for the Sun 1.408 g/cm^3). Unlike the Sun, Jupiter has a rocky core of heavy elements, and like other giant planets, it has no distinct solid surface.[100] Rapid rotation means that energy is actively being generated in the interiors of Jupiter, and part of the energy is being transferred to the rotational force of heavy elements.[*76)]

Figure 48 shows the internal structure of Jupiter. The mass of the core is estimated to be 12-45 times that of Earth or approximately 4-14% of the total mass of Jupiter, although its presence is known to be unclear.[101] The core is surrounded by a dense metallic hydrogen liquid. This liquid layer extends to about 78% of Jupiter's radius. Above the metallic hydrogen layer lies a transparent inner atmosphere of hydrogen in a supercritical fluid state. Jupiter's upper

---

*76) The Sun is actively generating energy inside, but its rotation speed is low.

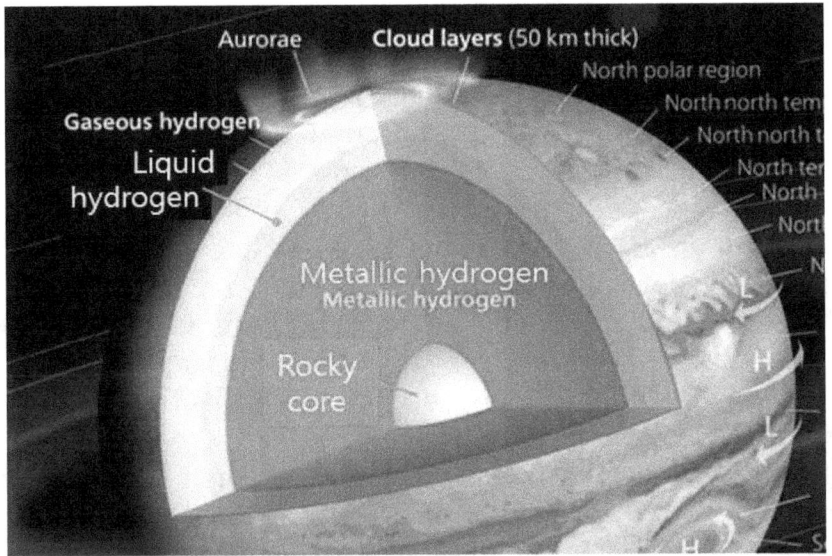

Figure 48. Inner structure of Jupiter (wikipedia.org).

atmosphere is of about 75% hydrogen, 24% helium, and 1% of other elements. Jupiter's interiors are relatively dense, with 71% hydrogen, 24% helium, and 5% other elements. The temperature and pressure inside Jupiter increase rapidly toward the center. At the surface, the pressure is 10 atm, the temperature is about 340 K. The temperature of the phase transition region where hydrogen is heated above the critical point and becomes metallic is estimated to be 10,000 K, and the pressure is 200 GPa (about 200,000 atm). The temperature at the core boundary is estimated to be 36,000 K and the pressure is approximately 3,000–4,500 GPa.[102]

Fate of Jupiter resembling the Sun

Current theories state that Jupiter's rocky core attracted hydrogen in the surround (by the action of gravity) to form Jupiter as it is now.[103] But this is not clear as the measurements of Jupiter's gravity are inaccurate.[104] It is also conjectured that early Jupiter, consisting only of hydrogen, was able to synthesize heavy elements through internal nuclear fusion and make a rocky core.

Stars including the Sun generate consistently energy by this fusion reaction to emanate to the outside, while synthesizing heavy elements, which undergo liquefaction and solidification to become a hard core. But when the rate of energy gain from the fusion process occurring inside of a star exceeds the critical rate, the star shall explodes and only its core will remain.[*77] As Jupiter is much smaller in size than the Sun, it will release the energy of nuclear fusion to the outside more slowly without this kind of explosion. The rocky core will grow and the apparent size of Jupiter will shrink. Indeed, Jupiter is decreasing in size by 2 cm a year.[105] Low energy release rates may produce liquid fuels such as methane as is found in Uranus. If the diameter is small, the pressure at the stellar center will also be low, making it difficult to produce heavy elements. If the internal pressure is relatively low, water or liquid fuel can also be formed.

---

*77) This is the birth process of a supernova and its neutron star, see Sections 3.5 and 3.6.

3.5. Supernova – explosion of the compressed energy

In the previous sections, we tried to interpret the internal structure and behavior of stars, the Sun, Earth, and Jupiter in terms of our vacuum paradigm. According to it, the core of a star or planet lies in a very high pressure state in response to the distortion of the surrounding solid vacuum, and that light elements are transformed into heavy ones through nuclear fusion. In this process, energy is compressed and concentrated, and some of it is released into space. Supernovae[*78] are "events" in which this process occurs explosively. The events shows how massive stars behave when the internal pressure is very high due to the stress of the surrounding solid vacuum. New, bright stars suddenly appear and then slowly fade away for weeks or months.

History of supernovae

Over the last millennium, three supernovae have been seen in our galaxy with naked eyes, and only five supernovae have been observed before the invention of telescope. The most recent witnessed in our galaxy is Kepler's supernova in 1604 (20,000 light-years from Earth, see Figure 49).[106] The first recorded supernova,

---

[*78] The term was first used in 1931 by Walter Baade (3.1893 – 6.1960, German astronomer) and Fritz Zwicky (2.1898 – 2.1974, Swiss astronomer).

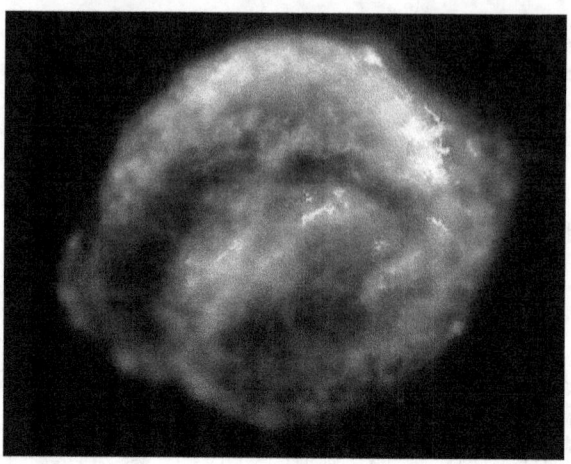

Figure 49. The remnant of Kepler's supernova. X-ray, infrared, optical image of SN 1604 (wikipedia.org).

SN 185, was discovered by Chinese astronomers in AD 185, and the largest supernova was SN 1006 (Figure 50), discovered by Chinese and Islamic astronomers in AD 1006.[107] Supernovae SN 1572 and SN 1604 (Kepler's supernova) are the last supernovae seen in our galaxy with naked eyes. Observing distant supernovae has become very common today, and hundreds are found each year. The brightest supernova so far is the ASASSN-15lh, first observed in 2015, with 570 billion times the solar brightness.[108] Its identity is not yet clear. A supernova releases explosively most of the stellar matter at the velocity of 30,000 km/s,[109] with a shock wave accompanied.[110] This wave sweeps away the surrounding gas dust and

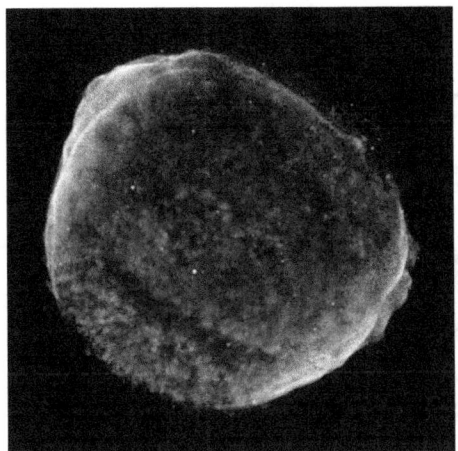

Figure 50. X-ray image of the remnant of SN 1006. Recorded as the brightest supernova until recently (wikipedia.org).

forms supernova remnants. Supernovae contribute to the generation of heavy elements, and shock waves can also mediate the formation of new stars.[111]

Type of supernovae

Supernovae vary in the degree of redshift[*79)]

---

*79) It is a phenomenon in which the wavelength of light emitted by an object is increased, causing it to appear red in the visible light region. As light-emitting objects move away from the viewer, the wavelength of light increases. The opposite of redshift is known as blueshift, which occurs when the object that emits light approaches the observer, or when light enters a gravitational field.

Figure 51. SN 1994D (bottom left), Type Ia supernova. It is brighter than the galaxy NGC 4526 next to it (wikipedia.org).

depending on the distance from us. The variation is reflected in the shape of the supernova's light curve (the change in the supernova brightness with time after birth). This is called the supernova spectrum and is used to study supernova physics and surrounding materials. Supernovae are categorized by the light curves and absorption lines appearing in the spectrum. If a supernova contains hydrogen absorption lines (the Ballmer series[*80]) of visible light)

---

*80) It describes the spectral line emissions of the hydrogen atom. The visible spectrum of light from hydrogen displays four wavelengths, 410 nm, 434 nm, 486 nm, and 656 nm, that correspond to emissions of photons by electrons in excited

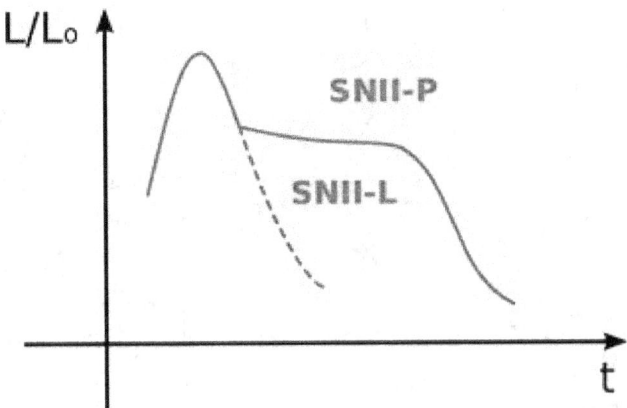

Figure 52. Supernovae are classified into Type II-P and Type II-L according to the shape of the light curve (SN = Supernova, wikipedia.org).

in the spectrum, it is Type II, if not it is Type I. A Type Ia supernova has additional ionized silicon (Si) absorption lines, while Types Ib and Ic supernovae have no such absorption lines. If a Type I supernova has helium absorption lines, it is Type Ib, if not it is Type Ic. A few Type Ic supernovae have a very smooth spectrum and overlapping absorption lines, indicating very high expansion rates. These are classified as Type Ic-BL.[112]

Supernovae are astronomical phenomena. The explosion of giant stars at the end of their life cycle

states transiting to the quantum state of the principal quantum number n = 2.

Figure 53. Birth of a supernova by thermal runaway
(Figures in wikipedia.org was modified).

has two theories. In the first case, a white dwarf absorbs materials from the neighboring partner and its temperature rises, triggering fusion and it becomes a supernova. In the second case, a sudden gravitational collapse leads to an abrupt release of the gravitational potential energy for a star to be a supernova.

Origin of supernovae - thermal runaway

The first theory is thermal runaway. When a white dwarf receives enough matter from the paring star as shown in Figure 53 and the temperature of the core rises, thermal fusion begins with carbon fusion. This is the theory of birth of Type Ia supernovae, though it's still unclear.[113] When white dwarfs composed of carbon-oxygen grow in size and reach the Chandrasekhar[*81)] limit (1.44 M$_\odot$, the solar mass), the electron degeneracy pressure[*82)] no longer withstands gravity and they collapse.[*83)] When the limit is approached, carbon begins to melt as the temperature and density in the stellar core rise. Within a few seconds, a lot of matter in the white dwarf does nuclear fusion and energy is released to an amount of $1\sim2\times10^{44}$ J, to give birth of a supernova.[114] At this moment, a shock wave of about 3% of the speed of light is generated and the brightness reaches -19.3 (5 billion times brighter than the Sun) in the absolute

---

*81) Subrahmanyan Chandrasekhar (10.1910 - 8.1995) Indian-US astrophysicist. Winner of the 1983 Nobel Prize in Physics for the research on the evolution of star structures. He found the Chandrasekhar limit of white dwarfs.

*82) It is a force generated by the Pauli exclusion principle that more than one electron cannot have the same position at the same time, which prevents white dwarfs from collapse.

*83) In the new paradigm, gravity itself does not exist. The gravitational force is the compressive stress exerted by the surrounding solid vacuum. The bigger the star, the greater the compressive internal stress, and above the critical point, the energy generated by nuclear fusion will be faster than the release rate, leading to an explosion, namely a supernova.

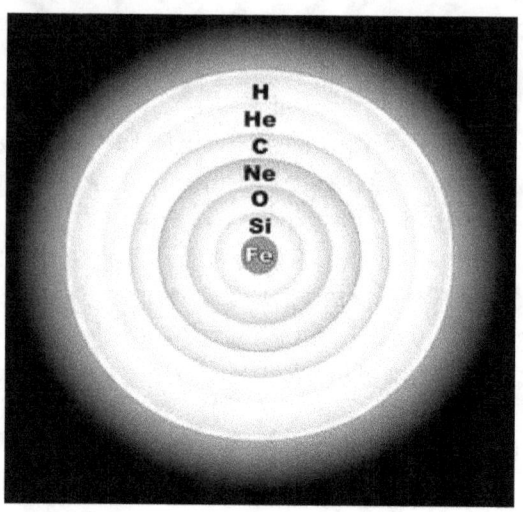

Figure 54. Onion-like structure in the core of massive stars (wikipedia.org).

magnitude.[*84] Referring to Figure 53, the bigger star first explodes in the binary system, forming a giant star, then reducing its orbital radius and becoming a white dwarf with carbon and oxygen as its main components.[115] The pairing star goes through the process of forming a red giant. The white dwarf absorbs matter from the red giant and increases its mass to a critical mass, and it eventually explodes to a supernova. In the light curve of Type Ia

*84) Apparent magnitude by assuming all stars are 10 parsecs(~32.6 light years) away. The absolute magnitude M is expressed as $M = m-5(\log_{10}d-1)$ as a function of the apparent magnitude m and the distance d.

supernovae, light is emitted during the degeneration of ^{56}Co from ^{56}Ni to ^{56}Fe.[116] In this process, the number of electron-proton pairs is reduced from 28 to 26. Surplus neutrons that are not paired with protons in the atom increased by 2.[*85]

Origin of supernovae - core collapse

The second theory is core collapse. Very massive stars cannot withstand their own gravity in the core, leading to core collapse. This is known to be the mechanism of all supernovae except Type Ia. When the core collapses, the outer layers can explode and become a supernova or the star may become a black hole or neutron star when the gravitational potential energy is insufficient (namely when the internal pressure is insufficient in our vacuum paradigm). Causes of core collapse include electron capture,[*86] exceeding Chandrasekhar limits, pair-instability; photodisintegration[*87].[117] If a massive star forms an iron core, it will exceed the Chandrasekhar mass and will collapse and evolve into a neutron star or black

---

[*85] In our vacuum paradigm, as discussed in Section 2.4, surplus neutrons have lower energy than other neutrons and play a very important role in the stability of heavy elements produced in the star.

[*86] Electron capture is a phenomenon where the nucleus of a neutral element with a high atomic number absorbs electrons inside, causing protons to become neutrons and release electron neutrinos.

[*87] A nucleus that absorbs gamma rays emits protons or neutrons and decay into light one.

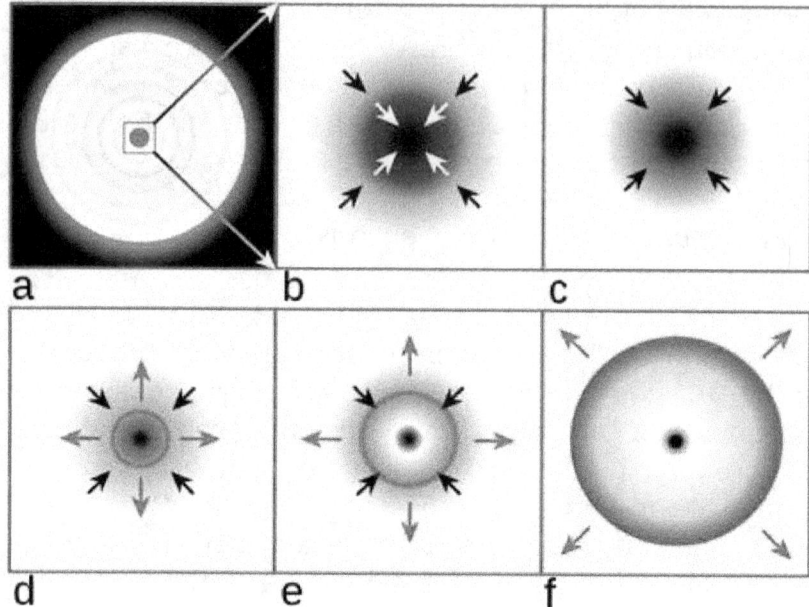

Figure 55. A evolved giant star (a) with the onion structure of Figure 54, where nuclear fusion occurs to form an iron core (b), begins to collapse when the Chandrasekhar (mass) limit is reached. The inner core is squeezed into neutrons (c) to release degenerate material to the outside (d) and form a shock wave (outer circle and outward arrow). The shock wave initially stops (e) but is invigorated again by some neutrino interaction. The surrounding material emanates (f), leaving only a remnant (wikipedia.org).

hole, because it cannot withstand electron degeneracy pressures.*88) Electron capture by Mg in the

---

*88) In terms of our vacuum paradigm, the greater the stellar mass, the higher the compressive stress in the core exerted by

degenerated O/Ne/Mg cores leads to gravitational collapse and explosive oxygen fusion. The formation of electron-positron pairs in the core after helium combustion induces a thermodynamic instability, leading to an explosion after the initial collapse, and eventually to give birth of a pair-instability supernova. When the core is big and hot enough, the resulting gamma rays induce photodisintegration and the core completely collapses.[*89]

If the initial mass is less than $1M_\odot$, the core is small and does not collapse and becomes a white dwarf. This white dwarf cools down without further fusion. Stars with more than $9M_\odot$ (or $12M_\odot$)[118] gradually burn heavy elements in the core, consume outer elements, and thus form an onion structure in Figure 54.[119] Most of these stars become core collapse supernovae, and are less bright and less visible than other giant supernovae. If the core collapses in a super-massive star with hydrogen in the outer shell, it becomes a Type II supernova.[120] This is the result of the rapid collapse of a heavy star and it is confirmed by hydrogen absorption lines in the emission spectrum, usually observed in the spirals of galaxies.

---

the surrounding solid vacuum. As a result, more neutrons are produced, releasing energy and becoming a lower energy neutron star or black hole.

[*89] When electrons and protons combine, gamma rays and low-energy neutrons are generated. If protons are positrons trapped in the lattice point of the solid vacuum, then the lattice points deprived of positrons is kind of neutron with only some vibrational energy remaining, resulting in higher density of neutrons in the core and eventually a neutron star.

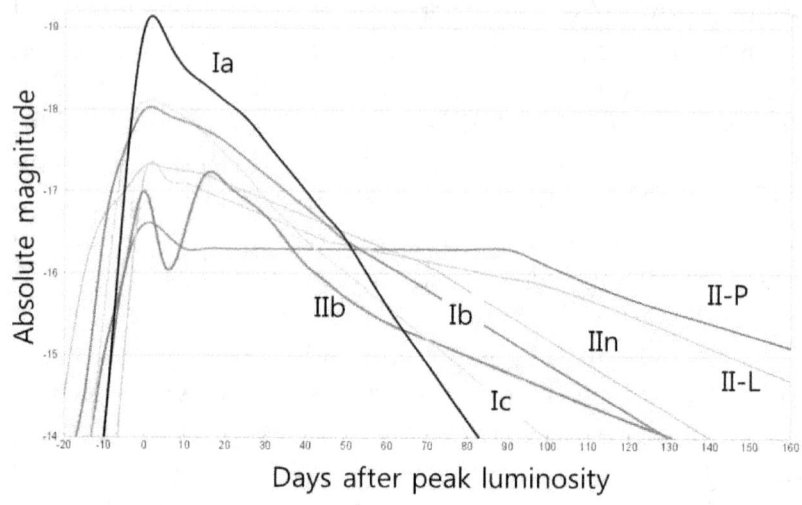

Figure 56. Light curves of supernovae (wikipedia.org).

Light curves of supernova

The supernova's light curve shows the process of energy release of a supernova. As shown in Figure 56, most of the supernova energy radiates instantaneously, resulting in a sharp increase in luminosity. The light curve shows the radiation characteristics of certain gases, first predicted in the late 1960s.[121] The gradual decay of luminosity in the light curve after the supernova birth is due to attenuation. Light comes mostly directly from the supernova, but it also comes indirectly from the remnants heated by the energy of erupted gases. ^{56}Ni emits gamma rays as it forms from ^{56}Co (847 keV)

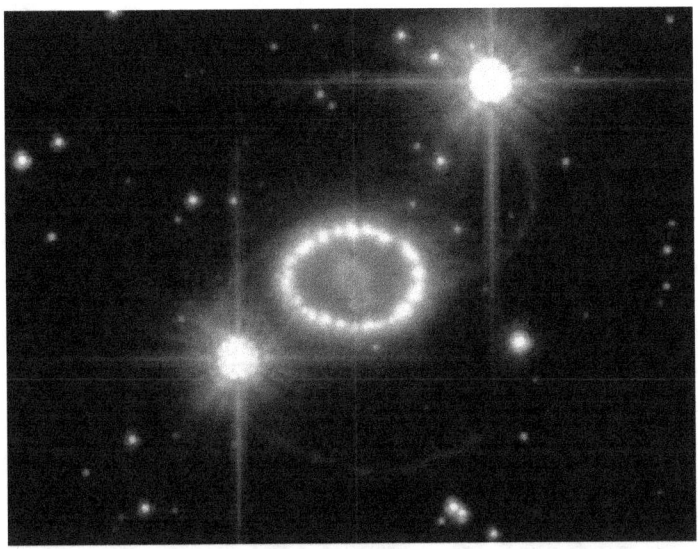

Figure 57. Remnant of supernova 1987A (central image, modulated image from wikipedia.org).

and decays to ^{56}Fe (1238 keV). Due to this energy the supernova shines for months.[122] In Type II supernova 1987A,[*90)] the peak in the light curve is reached when ^{56}Ni decays to ^{56}Co (6 days of half-life), followed by ^{56}Co to ^{56}Fe with the half-life of 77.3 days (see Figure 58). Analysis by cosmic gamma-ray spectroscopy identified some ^{56}Co and ^{57}Co gamma rays that were not absorbed by the remnants of SN 1987A, indicating

*90) Observed in 1987 near the Tarantula Nebula of the Large Magellanic Nebula. It is 16,8000 light years away. Neutrinos were observed from the supernova before visible light was observed. The release of neutrinos from supernovae was identified for the first time in history.

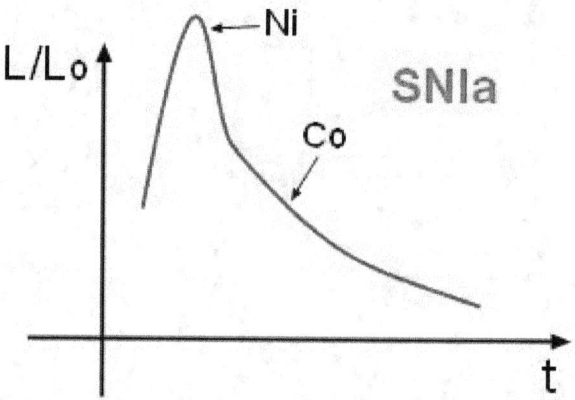

Figure 58. Light curve of a supernova indicative of ^{56}Ni and ^{56}Co radioactive decay (wikepedia.org).

that these two nuclei are the source of energy.[123]

Most of the light curves for Type Ia show a constant peak height and decrease in brightness relatively quickly. The luminosity is due to the radioactive decay of ^{56}Ni to ^{56}Co. The initial luminosity decreases rapidly as the captured electromagnetic waves disappear accompanied by the reduction of the effective size of the photosphere. After several months, the light curves become different in the extinction rate as the positron release is more pronounced from the remaining ^{56}Co. The light curves of Type Ib and Ic are similar to those of Type Ia, with only a slightly lower average of the maximum in luminosity. This is because the amount of ^{56}Ni is much less.

Type Ic supernovae, the brightest ones, are called hypernova or super-luminescent supernovae and have broader and bigger light curves. This kind of supernova yields a black hole rotating very fast, which in turn generates a very large jet to trigger explosive gamma rays at the outer edge of the star. If hydrogen is present in the erupted materials of a supergiant star, it is heated and ionized early in the collapse. Most Type II supernovae have elongated horizontal sections on the light curve, as ionized hydrogens recombine to shed visible light and become transparent. Type II-L supernovae has no such horizontal sections because the progenitor stars[91] had insufficient amount of hydrogen. Type IIn supernovae have an additional narrow spectrum generated from the dense surroundings. The light curve is very wide and sometimes very bright, so these suppernovae are called super-luminescent supernovae. The light curve indicates that the kinetic energy of the erupted materials is transformed into electromagnetic waves through the interaction with the dense material in the outer shell, being originated from the progenitor stars just before the explosion to supernovae.

Energy and matter erupted from supernovae

Supernovae are enormously bright due to the emission of electromagnetic waves, but their energy is only part of the story. This is especially true for

*91) A progenitor star is a star before becoming a supernova.

core collapse supernovae. Different types of supernovae emit various kinds of energy. The energy released from Type Ia supernovae, the explosions of white dwarfs, is consumed during the formation of heavy elements and transferred to the kinetic energy of erupted matter. The thermal runaway of carbon-oxygen white dwarfs induces nuclear fusion and almost all of the mass of the stars is released as kinetic energy. About half the solar mass is occupied by ^{56}Ni generated from silicon combustion. ^{56}Ni emits strong gamma rays via beta (β^+) decay*92) and turns into ^{56}Co. ^{56}Co goes through beta decay again and becomes stable ^{56}Fe. These two processes are the source of electromagnetic waves of Type Ia supernovae.124 As Type Ia supenovae generate kinetic energy and ^{56}Ni, they show low luminosity. Some of the energy amounting several times the solar mass is transferred to hydrogen to yield hydrogen ions. Hydrogen ions recombine with electrons, so that the rate of decrease in the luminosity is retarded.

In core collapse supernovae, enormous energy is released as neutrinos. More than 99% of neutrinos are released within minutes after the collapse. Core collapse supernovae appear faint than Type Ia supernovae, but the amount of released energy is much larger. The gravitational potential energy*93) of core collapse supernovae is converted into the kinetic energy, producing electron neutrinos in the first stage

---

*92) A process of positron emission. One proton in the nucleus releases a positron and a neutrino and turns into a neutron.
*93) It is the compressive energy of the solid vacuum.

of the collapse and then thermal neutrinos from the core of the remained extremely hot neutron star. The residues entering the black hole generated from core collapse supernovae form ultra-fast jets, which trigger gamma rays to erupt in one direction and to transfer to the surrounding materials. This is the story of highly luminous Type Ic hypernovae, for which gamma-ray explosion lasts for a long time.

Supernovae are the source of elements heavier than nitrogen, up to ^{34}S via so-called supernova nucleosynthesis. Elements between ^{36}Ar and ^{56}Ni are produced by silicon combustion,[*94] while elements heavier than iron are generated by colliding with neutrons during core collapse. This is called neutron capture and progresses rapidly (the r-process) or slowly (the s-process).[125] In the environment of high temperatures and neutron densities, neutrons are captured at very high rates, and the neutrons generated inside the highly unstable core are rapidly stabilized through beta decay. This process produces half of the isotopes heavier than iron, such as plutonium and uranium.[126]

Remnants of supernovae usually consist of hard matter such as neutron stars and of fast expanding matter shock waves. The shock waves are adiabatic, gradually cooled, and mix with the interstellar medium.[127] When the surrounding space is filled with matter other than hydrogen or helium after the birth of a supernova, the kinetic energy of the expanding

---

[*94] A nuclear fusion reaction that occurs within a short time in stars with the mass of 8-11 M☉.

Figure 59. Trilobite fossils presumed extinct in the Ordovician-Silurian period.

supernova remnants attract the surrounding matter to form another star.

It is estimated from the trace of short-lived radioisotopes that supernovae nearby to us have had a profound effect on the composition of the solar planets 4.5 billion years ago.[128] Under the influence of supernova gamma rays, nitrogen in the upper atmosphere is converted to nitrogen oxides, which in turn destroys the ozone layer and makes it be transparent to harmful ultra-violet rays. This was believed to be the reason for the Ordovician-Silurian extinction[*95)] and to be responsible for the death of

*95) The Ordovician begins with a small extinction of 488.3 million years ago and ends with a massive extinction of 443 million

nearly 60% of marine lives.[129] Traces of past supernovae could be detected from the metal isotopes in Earth's solid layers. ^{60}Fe was reported to have been concentrated in deep Atlantic rocks.[130] In 2009, nitride ions found in Antarctica ice were known to be due to SN 1006 and SN 1054. Their gamma rays raised nitrogen oxide concentrations and were trapped in ice. Type Ia supernovae are very dangerous when they are near Earth, because they can arise from binary white dwarfs and make a big influence on Earth's life. The closest candidate is Pegasus IK.[131]

Supernovae in the regime of our vacuum paradigm

How do supernovae come into being? As was noticed, there are two theories. The common thing is that when the gravitation force of a star exceeds its limit due to its own mass, rapid changes in the star occur via gravitational collapse. In our vacuum paradigm, there is no gravitational collapse because there is no gravity. The birth of a supernova is a process of releasing stress developed within a star exerted by the

---

years ago, resulting in the extinction of 60% of marine genera. Rocks of this period contain abundant fossils and also store fossil fuels. The Silurian is the shortest period of the Paleozoic Era that follows the Ordovician. It began at the end of the Ordovician and ended at the beginning of the Devonian 410 million years ago. The bottom of the Silurian is defined as the Ordovician-Sirurian extinction events, with about 60% of marine genera extinct. Oxygen was generated as a photosynthetic by-product of photosynthetic organisms, generating an ozone layer, and the first terrestrial organisms appeared.

surrounding solid vacuum, The internal stress increases infinitely in theory on going to the stellar center. Therefore, the center should be very high in pressure, and due to nuclear fusion, the temperature is also very high as the case with the Sun. A review of the solar system, including the Sun, confirmed that the core is composed of hot and heavy matter. Massive stars much larger than the Sun will have much higher pressures, and their cores can have very dense matter (high in the neutron density). Closer to the core center, the density increases rapidly, so an onion structure is inevitable as shown in Figure 54. The composition depends on the stellar size: the larger the star, the higher the central pressure and the higher the neutron density. As mentioned above in this section, neutrons are regarded as transients formed in the process of returning of hydrogen atoms to nothing, namely to the energy-free cold solid vacuum. Energy in hydrogen atoms is squeezed out via combining protons and electrons in the form of electromagnetic waves and neutrinos. In our new vacuum paradigm, surplus neutrons have less energy than paired ones with protons in heavy atoms. No wonder supernova remnants are neutron stars or black holes. They are ashes left on fire. As we noticed previously earlier in the behavior of the solar system, planets are consuming their remaining energy by rotation. Some energy remained in a neutron star, one of the remnants of a supernova, is the rotational kinetic energy. Pulsars, neutron stars that emits strong pulsating electromagnetic waves while rotating at very high speeds, were found recently. These neutron stars will also consume energy as they rotates and eventually turns into black holes. In the regime of our new vacuum paradigm, a black hole is just one of the central parts of a giant star in which energy is almost exhausted and thus appears to be "black". It is just an empty hole that absorbs matter and light like a black-body, when exposed to free space. It is not a monster that swallows anything including light.[96]

The higher the mass of the progenitor star, the more likely the supernova's remnant will become a black hole. This is because the pressure in the stellar core is higher and the contained energy is compressed more severely and exhausted more rapidly. Light elements turn into heavier ones, while energy is squeezed out (mostly in the form of gamma rays or neutrinos). When this process is not enough rapid, a star continuously dissipates heat as with the Sun. However, if the internal pressure is very high and thus the generation rate of energy is higher than the rate of energy dissipation, it will accumulate in the interiors of the star. The star will then explode without enduring this energy anymore at a certain critical point. A supernova is thus born. When a star explodes, there will be no remnants such as neutron stars or black holes if the explosion point is the stellar center. If it explodes at a certain distance from the center (if the core is already exhausted in energy, the core will be a transient state to a neutron star), it

---

*96) The paradox of black holes is discussed in detail in 3.7.

will radiate a shock wave outward and in respond to the explosion a momentary ultra high pressure will develop in the core. This causes the core material to undergo more and more energy compression and the remnant neutron star will turn into a black hole after a further energy release. In the light curve of Type Ia supernovae, light is emitted due to the decay process from ^{56}Ni to ^{56}Fe, which reflects this pressure change. Also, if a black hole first forms in the center of the progenitor star before it becomes a supernova, it will appear to collapse due to gravity. Black holes are similar to the cold solid vacuum in terms of energy stored, so if the shell of the black hole in the stellar center collapses, it will become a supernova and will not leave any remnants. Electron capture, pair instability, and photodisintegration, which are discussed as causes for core collapse, are all belonging to the process of releasing the compressed stress of the core developed by the surrounding distorted solid vacuum. The stellar core yields heavy elements in high pressure states, until it becomes a bulk of neutrons with low energy and eventually a black hole. Thus, the core of a neutron star will be composed of low-energy neutrons and the energy of neutrons will be higher toward the surface. In terms of matter, the inner core resembles an empty space, namely a black hole, surrounded by a hard shell rich in neutrons. This outer shell shall have an onion-like structure as similarly as shown in Figure 54. Whether a supernova is born or not will be determined by the amount of energy they accumulate in this shell.

3.6. Ultra-fast spinning neutron stars

What is the neutron star?

So far, it is believed that a neutron star is a star in the last stage of its life cycle where almost all of its energy is exhausted.[*97] A neutron star, being the remnant of a supernova, is known to come into being, as neutrons are produced by merging of electrons and protons in the process of so-called "gravitational collapse" at very high pressures. This neutron star eventually becomes a black hole allegedly by additional gravitational degeneracy.[132] In the regime of our vacuum paradigm, the cause of gravitational degeneracy is a very high stress field inside a star developed due to the distortion of the solid vacuum as much as the amount of (usually hydrogen) energy of the star. Under the influence of this stress, the energy of hydrogen (proton-electron) is compressed and squeezed, leaving only neutrons. The core of a massive star is concentrated in the number of neutrons, and a neutron star shows up as the remained neutron cluster of a supernova. A black hole is a state in which even the energy of the neutron star is exhausted and is free of energy similar to the cold solid vacuum. It is just a black-body with no energy and mass in itself.

A neutron star is estimated to be very dense, with a

[*97] It is not yet known whether this star is actually a neutron-only star or an electrically neutral that contains some protons and electrons.

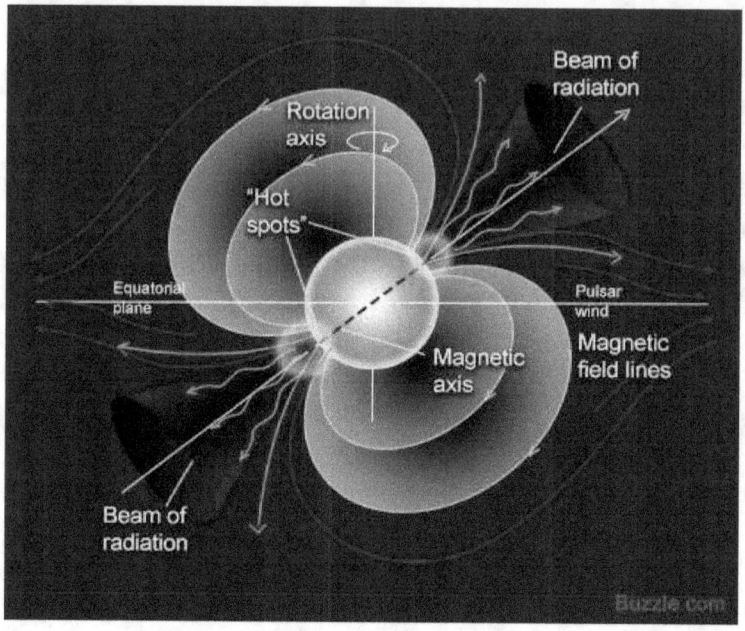

Figure 60. Pulsar electromagnetic waves are emitted from a neutron star, when the axis of rotation and the axis of magnetic field do not coincide (wikipedia.org).

radius of about 10 km, and its mass can be more than twice the solar mass. It is assumed that 100 million neutron stars has been produced by supernova explosions, but most are difficult to observe because of their age and low temperatures. The discovery of neutron stars is very recent. The observed neutron stars have very high surface temperatures of around 6×10^5 K,[133] and their magnetic field is 10^8~10^{15} times stronger than Earth's.[134] When the stellar core collapses, the angular momentum is conserved, so the rotation speed becomes high and rotates more than 100 times per second. The fastest rotating PSR J1748-2446ad do 716 revolutions per second, with a linear velocity of 24% of the speed of light.[135]

Discovery of neutron stars

The existence of neutron stars was first proposed in 1934,[136] and the following year it was predicted that the remaining core of a supernova would be a neutron star.[137] In 1967, Pacini[*98)] predicted that neutron stars would rotate and emit electromagnetic waves, if their magnetic fields are strong. In the same year, astronomers Hewish and Bell[*99)] confirmed the existence of the first neutron star by observing a electromagnetic pulse.[138] If the magnetic axis of a neutron star does not coincide with the axis of rotation, beams of electromagnetic radiation are emitted in all directions as shown in Figure 60. This neutron star is called a pulsar (pulsating radio star). Found in 1974, a pair of Pulsar PSR B1913+16 is a dual system that rotates around the center of gravity of the two neutron stars.[139] Neutron stars in a dual system attract and coalesce the surrounding material s,[*100)] in which x-ray becomes strong. In this process,

---

*98) Franco Pacini (5.1939 – 1.2012) Italian astrophysicist.
*99) In 1974, Antony Hewish (5.1924 -, British radio astronomer) won the Nobel Prize in Physics for the discovery of a pulsar, but his assistant Bell (Joslyn Bell, 1943.06 -, Northern Ireland astrophysicist) was excluded.
*100) The rotation of a star makes a two-dimensionally symmetric

old pulsars are also stimulated, and their rotational speeds increase. Two neutron stars can also cluster together, when gamma rays are released explosively in a short time or gravitational waves are emitted. According to General Relativity, when two large masses revolve in a short orbit, they emit gravitational waves, resulting in a further shorter orbital period. A pair of neutron stars emits strong gravitational waves just before the collision and merge when the acceleration is very high. This phenomenon was confirmed by the observation of PSR B1913+16, which indirectly demonstrated the existence of gravitational waves, and in 2017, a direct observation of gravitational waves due to the merging of a pair of neutron stars was reported.[140]

Gravitational waves from a dual neutron star

In General Relativity, gravity is a phenomenon originated from the distortion of spacetime by mass. The higher the mass density, the smaller the radius of curvature of the distorted spacetime on the surface of a spherical object. When an object moves in spacetime, the radius of curvature of nearby spacetime changes. As the object rotates, the change in the radius of curvature will change periodically and propagate outward in the form of waves. These waves can be called gravitational waves. As a gravitational

---

distortion of the solid vacuum as shown in Figure 32, so that an additional attracting force develops in the plane of rotation toward the center of rotation.

Figure 61. The crab nebula pulsar moves at 375 km/s against the nebula(wikipedia.org). Surrounding matter is concentrated due the distortion of the solid vacuum formed on the normal plane to the axis of rotation of the pulsar.

wave passes over an observer, nearby spacetime responds to this gravitational wave and thus time and distance are periodically shortened and lengthened in terms of General Relativity. This effect diminishes further away from the source.

This procedure can also be accessed in terms of the distortion of the solid vacuum in the regime of our vacuum paradigm. When two large neutron stars are close to each other as they rotate around their center of gravity, the solid vacuum becomes asymmetrical due to the two-dimensional distortion by the rotational motion in addition to the three dimensionally developed point-symmetrical distortion of the solid vacuum by their rest masses. Therefore, the distortion of the solid vacuum increases in the plane perpendicular to the axis of rotation, but changes periodically in that the period should be the orbital period of the two neutron stars. If a highly massive object, such as a neutron star, has a rapid rotational motion, we may imagine a situation that the distortion of the solid vacuum is so large and varies in the strength periodically that it propagates in a wave form. which may be called a gravitational wave. If a gravitational wave should exist, it is a periodical variation of the distortion of the solid vacuum. Mass is essentially the kinetic vibrational energy of the virtual vacuum lattice, and this mass determines the strength of the distortion of the surrounding solid vacuum.*101)

Three-dimensional point symmetry in the distortion of the solid vacuum induces an apparent attraction called gravity. Similarly, a rotational motion will cause a point-symmetrical distortion in the two-dimensional plane of rotation and cause asymmetric masses in the three-dimensional space. Therefore, it is like a mass-antimass fluctuation in that the strength of the distortion of the solid vacuum changes periodically due to the orbital motion of the massive neutron stars. For an observer on Earth, the combined mass of the two neutrons periodically increases and decreases due to the rotation of the neutron stars.

---

*101) In the regime of our vacuum paradigm, gravitational wave is nothing but matter wave propagate through the solid vacuum, so that its speed can be higher than the speed of light.

Structure of a neutron star

Neutron stars are known to consist almost exclusively of neutrons, and the gravitational force is expected to be very large. How was the mass of a neutron star calculated? Basically, it is calculated from the gravitational interference of the companion star based on Kepler's law, although there are many ways for i t.[141]

Since the companion star usually emits light from hydrogen on the surface, and the surface gravity is calculated by analyzing the hydrogen spectrum (regarding to the variation of its wavelength). The wavelength of light is affected by the gravitational field, that is, by the distortion of the solid vacuum. From the calculated value of gravity, the mass is determined based on the assumption that General Relativity holds.[*102)] In fact, the gravitational interference is, in the regime of our vacuum paradigm, the distortion of the solid vacuum by the companion star, which also varies when there is a rotational motion, as mentioned earlier. The rotation of the neutron star means that the calculated mass determined by analyzing the hydrogen spectrum depends on the position of the rotation axis.

If a neutron star has the structure as shown in Figure 62, the interiors may consist only of neutrons.

---

*102) Surface gravitational acceleration $g \equiv GM_{WD}/R_{WD}$, G is Newton's gravitational constant, WD refers to the "white dwarf", the companion of a neutron star. M and R represent the mass and radius, respectively.

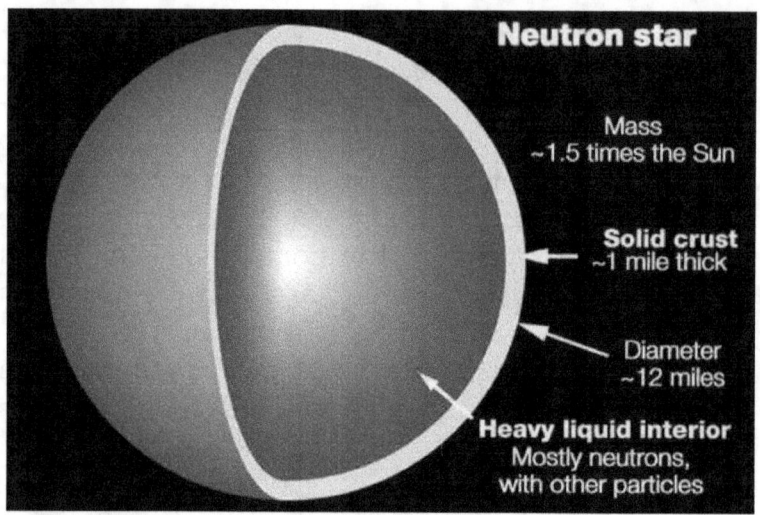

Figure 62. The structure of a neutron star (wikipedia.org).

Depending on the amount of energy of neutrons, it can be regarded as matter or as a part of the cold solid vacuum. In the early stages of the formation of a neutron star, its energy is so high that it distorts the solid vacuum due to its high rotational speed. However, when energy is exhausted, the neutron star distorts the solid vacuum so small that the two-dimensional gravity should be very small, and thus a much less mass for the neutron star would be obtained.

Neutron stars in our vacuum paradigm

As we can assume from the internal structure of

Earth, the stellar center would be very high in pressure, concentrated in neutrons. When neutrons are highly pressurized due to the surrounding distorted solid vacuum, they can have a very large amount of energy in their small volume. If there is no outer solid shells for neutron stars as contrarily as with Earth, the energy will be released outwards rapidly (if the energy of neutrons is very large, the neutrons will be converted to protons and electrons through beta decay) and will eventually become a part of the cold solid vacuum. An energetic neutron star may be regraded as a huge atom. Heavy elements (e.g. uranium: atomic number 91) formed in the core of giant stars are richer in the number of surplus neutrons than light elements. From this fact it can be inferred that the interiors of a neutron star is composed of a very large number of surplus neutrons and the outer layers has a structure similar to ordinary matter composed of protons and electrons (similar to heavy elements). The strong magnetic field of a neutron star further means that the flow of electrons or protons (or positrons) that make up the outer shell of the neutron star is strong. The inside is filled with neutrons whose physical properties are different depending on the amount of remained energy. The closer to the center, the property of neutrons will be more similar to that of the cold solid vacuum.

If a neutron star is left after a supernova explosion, it could be composed of dense matter, as seen in the internal structure of Earth. However, when a massive star explodes to a supernova, the stellar diameter will be reduced drastically, through which the central stress (by the surrounding solid vacuum) will be almost eliminated. The energy condensed in the stellar core will inevitably be released as it becomes a neutron star, and some energy remains in the form of rotational energy with a very high speed and a strong magnetic field. A new born neutron star from a supernova consumes energy as it emits electromagnetic waves in the form of pulses, and the neutron star will eventually fade away to a black hole absolutely free of energy.

3.7. Paradox of black holes

Black holes are believed to be the destination of neutron stars, and gravity is so strong that even light cannot escape therefrom.[142] General Relativity suggests that very dense materials distort spacetime so strong as to create black holes.[143] In our vacuum paradigm, however, black hols are nothing but empty spaces, namely parts of the cold solid vacuum free of energy.

Birth and behavior of black holes

Black holes were theoretically born from the singularity included in the first exact solution of Einstein's equations (1.9), derived by Schwarzschild in 1916. Eq. (3.1) is the solution of Schwarzschild, which will be discussed in detail later in this section. It is

Figure 63. A black hole simulated. The gravitational lensing effect severely distorts the appearance of galaxies around the black hole (wikipedia.org).

stated, when a black hole is formed, it absorbs surrounding matter and grow. Extra-large black holes with millions of the solar mass can allegedly be formed by absorbing other stars or black holes. The mass of the black hole, believed to exist in the galactic center, can be obtained from the orbital properties of a star orbiting its surrounding (e.g. Sagittarius A*). The mass is estimated to be 4.3 million times the solar mass. Recently, LIGO[*103]

*103) Laser Interference Gravitational-Wave Observatory, Currently located in the two States, Washington and Louisiana, for the observation of gravitational waves.

Figure 64. Conceptual view of x-rays emitted from a black hole (wikipedia.org).

indirectly proved the existence of black holes by directly detecting gravitational waves resulting from the collision and fusion of a pair of black holes.[144]

At a certain distance from the center of a black hole is the so-called event horizon, at which any object must have the speed higher than that of light to escape from the black hole. That is, even light cannot overcome the gravitational field of a black hole at the event horizon, but its existence has not been confirmed yet. Black holes are sometimes treated as ideal black bodies that do not reflect light at all.[145] Our paradigm also treats black holes just like black bodies.

According to quantum field theory in curved spacetime, the event horizon emits Hawking[*104)]

---

*104) Stephen William Hawking (1.1942 ~ 3.2018) British theoretical radiation. Its temperature is the same as that of the spectrum from a black body, but it is too small to be measured. Even in very small amounts, the release of radiation heat will cause black holes to gradually die out, and thus small black holes are expected to disappear quickly.[146]

Giant black holes are believed to shine brightly on x-rays. Some x-rays come from the surrounding disk and some come from the corona of the disk. Figure 64 conceptually illustrates this phenomenon.[147] When the corona is inclined to the black hole, the black hole has a strong attraction to the corona x-ray. As a result, x-rays are severely distorted as shown in Figure 65. According to the law of conservation of angular momentum, gases falling into the gravitational well developed around a very massive object, such as a black hole, typically have a disk-shaped structure(it is called an accretion disk). This disk-shaped gaseous lump transfers angular momentum out via the internal friction and falls faster into the center. At this moment, potential energy is released and the gas temperature increases.[148] Due to the friction the temperature increases so high that electromagnetic waves in the form of x-rays up to 40% of the mass of the gaseous materials are emitted.[149] (In nuclear fusion, only about 0.7% of the mass is emitted.) In many cases, very fast jets expelled from the center of the contracting accretion disk are accompanied. Active galactic nuclei and quasars*[105) can be

---

physicist. Born on the 300th anniversary of Galilee's death. He proposed Hawking radiation from quantum field theory.

interpreted as the accretion disks of supermassive black holes.[150] In November 2011, a disk-shaped quasar around a giant black hole was first observed directly.[151]

Controversy over black holes

In 1971, Hawking showed that the total area of the event horizon of a black hole does not decrease regardless of the collision or merging of two black holes.[152] This is now known as the second law of black hole dynamics, and is very similar to the second law of thermodynamics, in which the total entropy of a system does not decrease. At 0 K, a black hole assumes zero entropy. When entropy-containing matter enters the black hole at 0 K, it means that the entropy of the universe is reduced, which violates the second law of thermodynamics. Because of this, Bekenstein[*106)] assumed that black holes have entropy and will be proportional to the event horizon area.[153]

Hawking proved based on quantum field theory that

---

*105) Quasars (quasi-stellar objects) are giant luminous bodies that are formed by the generated energy when black holes swallow up surrounding materials. At the center of a quasar is a very massive black hole, one billion times the solar mass, surrounded by a disk, the materials of which are rotating and falling into the black hole, where the potential energy is released as electromagnetic waves.

*106) Jacob David Bekenstein (5.1947 – 8.2015). Mexican-born Israeli-American theoretical physicist. He played an important role in establishing the basis of black hole thermodynamics.

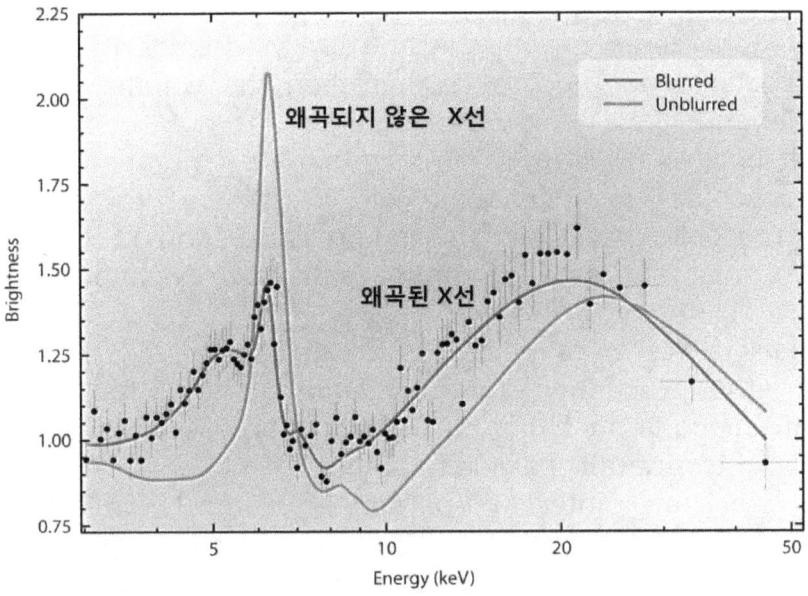

Figure 65. Distortion of x-rays near a black hole (NuSTAR, NASA's Nuclear Spectroscopic Telescope Array; 12 August 2014).

the radiation from a black hole, i.e Hawking radiation, occurs at a constant temperature. This radiation appears to violate the second law of black hole dynamics because it takes energy away from the black hole and thus the black hole has to contract. However, this radiation does not violate the thermodynamics, because the general assumption that the sum of entropy of matter around a black hole and the area of the event horizon are always increasing can be proved, though the radiation takes energy and entropy away. What is strange about this radiation theory is that the entropy of a black hole is proportional to the area rather than the volume. The explanation by General Relativity is not enough. Statistically, entropy represents the number of microscopic arrangements of a system for the macroscopic properties (mass, pressure, and charge). Even quantum gravity cannot satisfactorily interpret it.[154]

Black holes were born based on General Relativity. But the existence is challenged due to the contradictions such as "the information loss paradox" and "the firewall paradox". Since black holes have only a few internal parameters, most of the information of matter is lost in black holes. Only total mass, charge and angular momentum are preserved. As a black hole evaporates slowly, the information of Hawking radiation is lost forever. In quantum mechanics, the information loss violates the laws of conservation of energy and of probability[*107)]. This is the paradox of information loss over black holes. There are evidences that this problem can be circumvented by quantum gravity in that information and probabilities are indeed preserved.[155] However, Einstein's equivalence principle that gravitational mass and inertial mass are the same, the law of conservation of probability, or quantum field theory should be abandoned due to the firewall paradox. Quantum field theory states that two entangled

---

*107) The sum of the probabilities of all events that can occur in a quantum system is always 1, the unitarity law,

particles are involved in one emission of Hawking radiation. One particle is emitted as a quantum of Hawking radiation and the other is swallowed by the black hole. If a black hole formed at a time in the past evaporates completely at some time in the future, only the limited information is emitted by Hawking radiation. Assume that more than half of the information has already been released. The emitted particles are then entangled with all the Hawking radiation previously emitted by the black hole. This is the firewall paradox. According to the monogamy of entanglement principle, the emitting particles cannot entangle completely with two independent systems simultaneously. The emitted particles are entangled with the incoming particles and simultaneously independently entangled with the past Hawking radiation.[156]

Paradox of black holes

Black holes were initially believed to absorb everything including light, and never release it again due to their enormous gravity. At one time, however, Hawking insisted that even black holes emit faint light, and even further it was speculated that super massive black holes would have the density of water and very weak in the event horizon.[157] It is also stated that black holes evaporate with Hawking radiation and eventually die out.[158] In the establishment of black hole theories, some contradictions were argued, such as the violation of the second law of thermodynamics

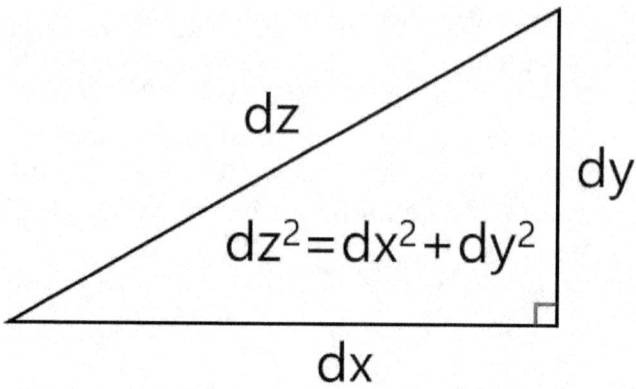

Figure 66. Pythagoras' Theorem.

mentioned above. There seems to no precise physics for black holes yet. The concept of gravity or General Relativity cannot satisfactorily interpret the behavior of black holes. At the center of the arguments lies the so-called "event horizon", from which even light cannot escape. The concept of the event horizon was originated from the Schwarzschild metric derived as a solution to Einstein's field equations.[159] The metric is give as

$$ds^2 = \left(1 - \frac{r_i}{r}\right)c^2 dt^2 - \frac{dr^2}{\left(1 - \frac{r_i}{r}\right)} - r^2 d\theta^2 - r^2 \sin^2\theta d\phi^2 \quad \text{---} \quad (3.1).$$

Eq. (3.1) appears to be quite complex, but it just means the shortest distance (shortcut or geodesic) between two points in four-dimensional spacetime. The shortest distance between two points in two dimensional space is represented by Pythagoras' theorem, as shown in Figure 66. In a three-dimensional Cartesian coordinate system it is given as

$$ds^2 = dx^2 + dy^2 + dz^2 \quad \text{---} \quad (3.2).$$

Eq. (3.2) can be extended to four-dimensional Minkowski's space by introducing Einstein's concept of spacetime. Eq. (3.1) is an expression in the polar coordinate system, where c is the speed of light, t is the time, and θ and φ are the azimuth and altitude, respectively. Referring to this metric, two singularities are noticed. One is at $r = 0$ (the central point) and the other is at $r = r_i$. If $r = 0$, the first term in Eq. (3.1) is infinite; if $r = r_i$, the second term is infinite, where r_i is called the "Schwarzschild radius" contained in Eq. (2.14).

Originally the solution to Einstein's equations, Eq. (3.1), was not relevant to black holes. There has been much discussion about the physical meaning of the Schwarzschild radius, and eventually $r = r_i$ has resulted in the "event horizon".[160] In Eq. (2.14), if $r = r_i$, the escape velocity from any spherical object of mass M is c, so that even light known to be the fastest in the universe cannot escape from this event horizon. Thus "mathematical black holes" or Schwarzschild black holes were born. In fact, the Schwarzschild metric of Eq. (3.1) is also a weak field approximation. Rewriting Eq. (2.14) we have

$$\frac{1}{2}mv^2 = \frac{GMm}{r} \quad \text{--- (3.3).}$$

Referring to this equation, the kinetic energy of a solar planet in the left-hand side is finite because the velocity does not exceed the speed of light, but the potential energy is infinite at $r = 0$. Namely, Eq. (3.3) is not equal in a strict sense. In our vacuum paradigm, gravitational potential energy and kinetic energy have the same origin, the distortion of the solid vacuum. The term expressed as kinetic energy on the left side is an approximation resulting from the assumption that the velocity of an object is very low compared to the speed of light. The distortion energy of the solid vacuum due to an actual motion can be obtained from the difference between the relativistic mass and the rest mass, as shown in Eq. (2.12). An exact description for Eq. (3.3) is then

$$(\gamma-1)mc^2 = \frac{GMm}{r} \quad \text{--- (3.4).}$$

Rearranging Eq. (3.4) for v, we have

$$v = \frac{c\sqrt{r_i(4r+r_i)}}{2r+r_i} \quad \text{--- (3.5).}$$

In this equation, if $r = r_i$, $v = c\sqrt{5/3} \approx 0.754$, and $v = c$ only at $r = 0$. There is no "the event horizon" based

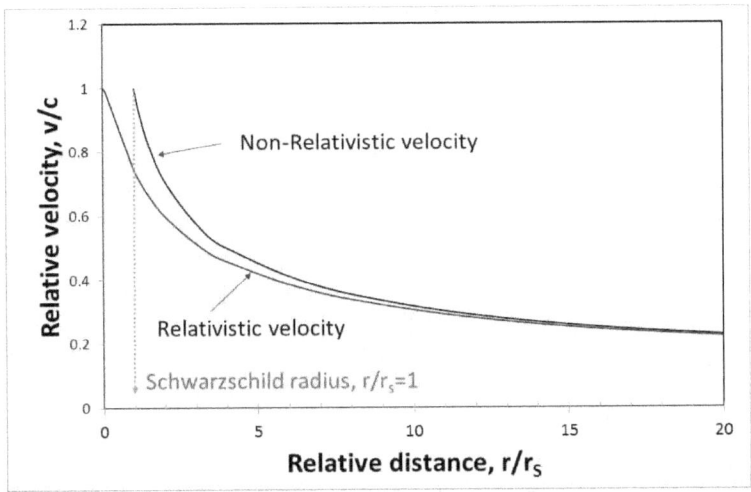

Figure 67. Comparison of the non-relativistic velocity yielding a singularity at the Schwarzschild radius with the velocity considered in our vacuum paradigm (relativistic velocity).

on the Schwarzschild radius, and there are no "mathematical" Schwarzschild black holes. Figure 67 compares the velocity obtained from Eq. (3.3) and Eq. (3.4) as a function of distance. In the regime of our vacuum paradigm, v gradually increases up to the speed of light as it approaches the stellar center, not goes through the singularity at $r = r_j$.

New interpretation of black holes

Let us interpret the behavior of currently known black holes in the regime of our vacuum paradigm. First of

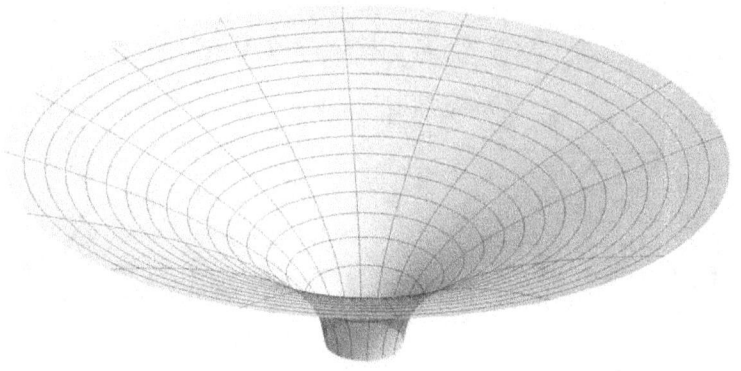

Figure 68. Flamm's paraboloid for Schwarzschild metric. Equivalent to the non-relativistic velocity as a function of distance in Figure 67.

all, if black holes are evolved from neutron stars, they should be very dense. This means that they are relatively small, and indeed, there are many very small black holes. In our new vacuum paradigm, a highly massive star that has a very small volume like a black hole will exert stress as much as the energy (mass) on the surrounding solid vacuum. But if a neutron star is already almost exhausted in energy, the black hole evolved from the neutron star will make a tiny influence on the solid vacuum. This is because the volume and energy are smaller than those of the same neutron star, and the property gets more to that of the cold solid vacuum. While emitting a slight amount of light (Hawking radiation), this black hole will eventually fade away to be a part of the cold solid vacuum free of energy, namely it dissipates or evaporates. It also means that the gravitational force is weakened in terms of Newtonian mechanics (in fact, the distortion in the solid vacuum is weakened), and therefore the mass calculated in the regime of General Relativity or Newton's gravity theory is also reduced. Of course, the mass density is as well reduced. The low density of very large black holes can be understood as such. If the origin of gravity (in a two-dimensional plane) is the asymmetric distortion of the solid vacuum induced by the rotation of the object, the energy reduction of a black hole is nothing but the reduction of the rotational energy, so that the size and density of a black hole decreases as its rotational energy is released in the form of Hawking radiation.

The distortion of the surrounding solid vacuum by an spherical object (such as the Sun) exerts in return a stress concentrating at the center of the object, and thus the internal stress of the object inevitably become singular at the center as shown in Figure 22. As we have seen in the cases of the Sun, Earth, and Jupiter, the closer to the core, the higher the density, temperature, and pressure, and the more energy is squeezed. If the energy in the core is exhausted at high temperatures and pressures, this region will gradually become a part of the cold solid vacuum it self. This is the black hole in our vacuum paradigm. This black hole is not a monster that absorbs all of its surroundings because of its very high gravity, but is just the solid vacuum itself free of energy. Our black hole is a part of the cold solid vacuum where almost all the energy is exhausted. In terms of energy and matter, it is just a literally "black" hole. Just as water is sucked into a hole, mass and energy around a black hole will be sucked into it. However, because the capacity of a black hole to hold energy or matter is limited, the hole will be filled up and therefore disappear. This process is very natural and does not violate any of the currently known laws of physics. The disappearance of black holes is a process of increasing the entropy of the universe, and there is no information loss and no firewall contradiction in the regime of our vacuum paradigm.

IV. For the new universe

Black holes in our new universe are not monsters that suck even light because of their enormous mass, but are just parts of the cold solid vacuum where the energy of matter is fully exhausted. Surrounding matter and light flow into this empty space naturally, and the inflows flow out again. The observations of the International Center for Radio Astronomy Research (ICRAR), published in the latest issue of Nature on April 30, 2019, illustrate the very nature of black holes. It is one of the results of inverting the conventional concept of black holes that nothing can escape.

4.1. Cosmic Microwave Background

In our vacuum paradigm, the energy of matter is essentially vibrational kinetic energy of the solid vacuum, and this vibration creates the so-called gravitational attraction force. This attractive force is just an apparent one originated from the distortion of the solid vacuum, which in turn pressurizes matter. Therefore, the stellar center lies in a state of very high temperatures and pressures and the energy of star (mostly hydrogen energy) is squeezed outward slowly or explosively in the form of supernova. Where does the released energy ultimately go? If the universe made of the solid vacuum is infinite, the emitted energy will spread all over the universe in the form of waves (electromagnetic waves or neutrinos) and the nearby universe will cool off. If the universe is finite, the released energy will be preserved and serve to raise the temperature of our universe. It is the Cosmic Microwave Background (CMB) that the released energy remains in the solid vacuum. It is a faint radiation that forms the background of the universe, filling the entire universe as shown in Figure 69 (Is the universe finite?). Highly sensitive radio telescopes show almost uniform noises. The wavelength is the most intense in the microwave range. The existence of the CMB strongly suggest that the interstellar space of the universe is not empty. However, current astrophysics states that electromagnetic waves generated in the early universe after the Big Bang are left behind, and the CMB is

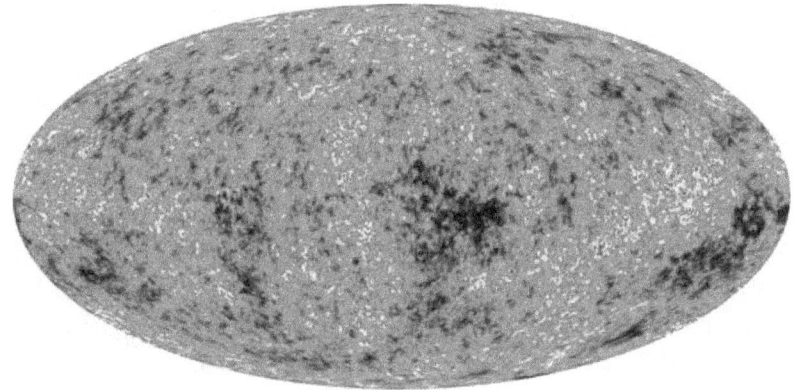

Figure 69. Cosmic image obtained from the Wilkinson Microwave Aniostropy Probe (WMAP) data for nine years (wikipedia.org).

recognized as a monumental discovery that confirms the existence of the Big Bang.

Origin of the CMB

The CMB is a black-body radiation at 2.72548 K. When the radiation energy is differentiated by frequency (dE_v/dv), the maximum frequency is 160.23 GHz and is in the microwave range (~6.6×10^{-4} eV in energy). When it is differentiated by wavelength ($dE_\lambda/d\lambda$), the maximum wavelength is 1.063 mm. The radiation is almost uniform in all directions but nonuniform by 1 ppm and by 18 μK in temperature.[161] In particular, some anisotropy is measured differently depending on the region in the spectral radiance viewed from different angles. The Big Bang model represented by the Λ-CDM model is allegedly most likely to explain this anisotropy.

According to the Λ-CDM model,[*108] the universe expands exponentially as shown in Figure 82 and in 10^{-37} seconds after the Big Bang, almost all the inhomogeneity disappeared,[162] and the remaining inhomogeneity led to a further expansion by quantum fluctuation[*109].[163] Before the formation of stars and planets (after 10^{-6} seconds of the Big Bang), the early universe was small, much hotter, and filled with a uniform plasma with interacting photons, electrons, and baryons.[*110] As the universe adiabatically expanded and cooled down, electrons and protons combined to form hydrogen atoms. These atoms could not absorb radiation heat and the universe became transparent. This period is called the recombination epoch, and the universe was about 3,000 K in temperature and 379,000 years old.[164] Since photons

---

*108) The Λ-CDM model is theorized by introducing the cosmological constant Λ, dark energy, and cold dark matter (CDM) to the Big Bang hypothesis. In this model, dark energy and Λ are identified. It was born based on General Relativity in the 1990s and is stated to best explain the structure of the CMB.

*109) Quantum fluctuation (or vacuum fluctuation) is a momentary change in energy at a point in space based on the uncertainty principle of Heisenberg (Werner Heisenberg, 12.1901 – 2.1976, German physicist). Virtual particle-antiparticle pairs are then generated and can be measured indirectly. For example, an effective charge whose electron charge is different from the actual value is measured.

*110) Particles made up of three quarks. Protons and neutrons are typical baryons.

Figure 70. Holmdel Horn antenna for observing the CMB (Penzias and Wilson).

do not react with hydrogen atoms, they were free to move through the space and separated from matter.[165] The period when light moves freely in the space of no electrons and protons is called the photon decoupling period, when the temperature of photon was lowered to 2.726 K, which is the same as that of the black-body radiation, and the wavelength increased with the expansion of the universe. So the original photons emitted during the photon decoupling epoch are said to make up the current CMB.[166] According to the Λ-CDM model, the radiation currently measured comes from a spherical surface called the surface of the last scatter. This radiation shows a group of positions in the space where the separation took place

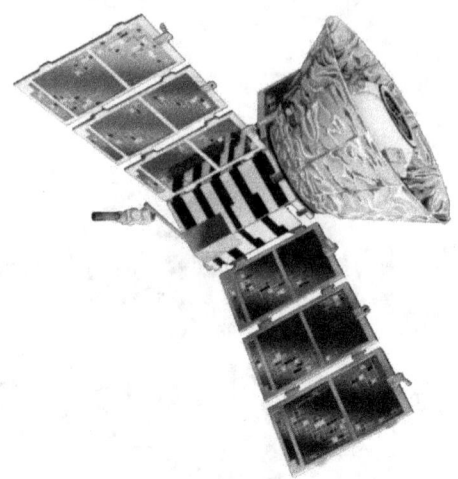

Figure 71. NASA's Cosmic Background Explorer COBE 66 operated by NASA to observe the CMB from 1989 to 1993.

and indicates that photons at that time and location have reached current observers.[167] Most of the cosmic radiation energy is from the CMB and is about 6×10^{-5} of the whole cosmic density.[168]

Discovery and observation of the CMB

How was the CMB discovered and got noticed? In the 1940s, some astronomers were interested in the temperature of the universe. In 1948, Alpher[*111)] and Herman[*112)] predicted that the temperature was 5 K,[169]

---

*111) Ralph Alpher (2.1921 - 8.2007), US astronomer.

but corrected to 28 K thereafter. This high calculation was due to a incorrect value of the Hubble constant, *113) and there was no accurate concept for the CMB. While it attracted little attention later on, it was newly predicted as the CMB in early 1960s,[170] and was first observed in 1964 by Penzias*114) and Wilson*115) using an antenna shown in Figure 70.[171] The temperature of the CMB was 4.2 K. In 1990, NASA observed the CMB with the COBE satellite FIRAS(Far-InfraRed Absolute Spectrometer), as shown in Figure 71, identified it as a black-body spectrum as shown in Figure 72. As a result, the CMB began to be recognized as a residue of the Big Bang, which cannot be explained by the steady state model.*116) In 1992, the COBE satellite measured the CMB more accurately with a Differential Microwave Radiometer, and found some anisotropy in the spectrum. Any effects of cosmic strings on this anisotropy could not be confirmed, thus superstring theory*117) that cosmic strings are the main

---

*112) Robert Herman (8.1914 ~ 2.1997), US astronomer.

*113) Constant according to the Hubble's law that the redshift from a distant galaxy is proportional to the distance. Hubble's observations in 1929 estimated 500 (km/c)/Mpc, now around 70 (km/c)/Mpc. Mpc=megaparsec, parsec~ 3.26 light years.

*114) Arno Allan Penzias (4.1933 -), US astronomer from Germany.

*115) Robert Woodrow Wilson (1.1936 ~), US astronomer.

*116) It is one of the theories of the universe that was popular in the mid-20th century. Since the expanding universe produces matter continuously, the density of matter is constant, so the celestial law of the universe is always the same anytime, anywhere. This theory became obsolete because it was unable to describe the current structure of the universe well.

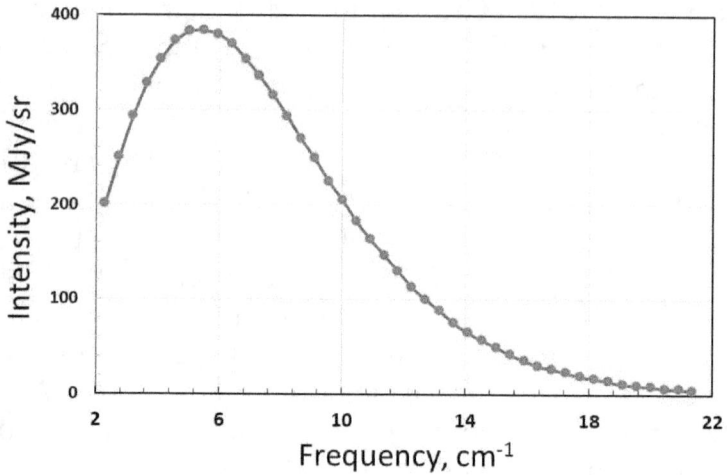

Figure 72. CMB spectrum from COBE. Red points are measured data. The solid line is black body spectrum.

components of the cosmos was eliminated. It was confirmed that the expansion is the correct theory for the formation of the current cosmic structure.[172]

NASA observed anisotropy in the CMB on a large scale from 1989 to 1996 with the COBE satellite. Since then, various observations have been made, supporting the current theory of the expansion. Throughout the 1990s and 2000s, the universe was regarded to be flat. In 2003 WMAP analyzed the spectrum signals in detail with an accuracy less than 1° for the first time and presented large scale angle

*117) Superstring theory is a theory that explains all the particles and fundamental forces that exist in nature in terms of the vibration of super-symmetric strings.

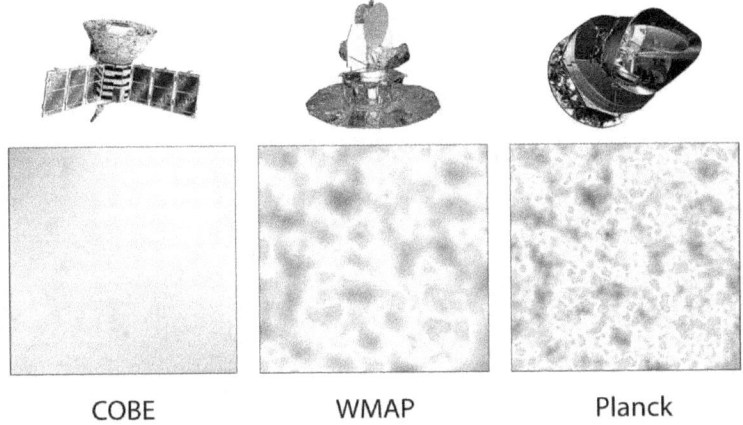

COBE　　　　　WMAP　　　　　Planck

Figure 73. Comparison of the CMB (March 21, 2013).

fluctuations in the CMB.[173] The European Space Agency's Plank spacecraft,[*118] equipped with more precise instruments, observed the CMB from May 2009 to October 2013 in space. In March 2013, Planck's space detector presented a CMB map, which showed slight temperature fluctuations in the deep sky when the universe was 370,000 years old.[174] These fluctuations reflect the waves that generated 10^{-30} seconds after the Big Bang. These waves gave rise to the cosmic scale webs of today's massive galaxies and dark matter. As WMAP data became more sophisticated, abnormal phenomena of the CMB, such

*118) A space station observatory operated by the European Space Agency (ESA) from 2009 to 2013, where the CMB anisotropy was measured in the microwave and infrared ranges using a very high precision instrument.

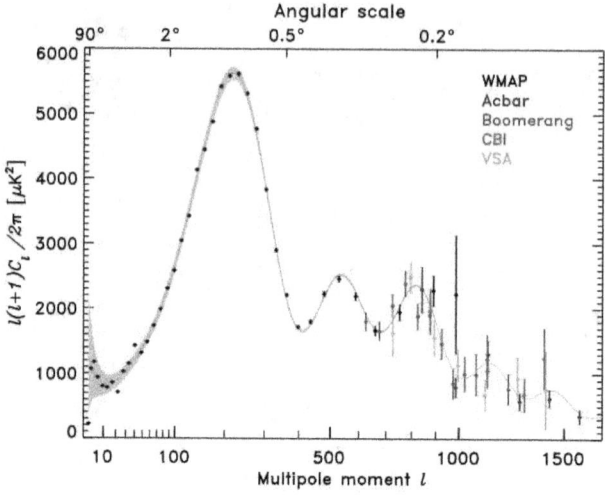

Figure 74. Relationship between the CMB temperature anisotropy and the energy density spectrum based on multipole moment (wikipedia.org).

as the large scale anisotropy, abnormal alignment, and non-Gaussian distribution were analyzed.[175] This precise analysis reveals anomalies that cannot be explained by the Big Bang theory. The same phenomena were observed with the Planck space telescope,[176] which some scientists interpret as a signal for new physics.[177]

Current analysis of the anisotropy in the CMB

The anisotropic structure of the CMB is known to be due to the effects of acoustic oscillations and photon diffusion damping.*119) Acoustic oscillations come from the friction between photons and baryon (matter) plasmas in the early universe. Light tends to erase anisotropy, but baryons move at much lower speeds than light and aggregate under the influence of gravity. These two effects cause acoustic vibrations, forming a structure in the CMB with picks as shown in Figure 74. From the angle of the first pick the curvature of the shape of the universe can be obtained, and the second one determines the reduced baryon density.[178] Information on the density of dark matter can be allegedly obtained from the third peak. [179] Dark matter is discussed in detail in the next section.

The location of the picks also provides information about the origin of primordial density perturbations, being the adiabatic and isocurvature, usually a mixture of the both. The adiabatic density perturbation is equal to the excess density ratio of individual particles. Namely, in some places the density of baryons is 1% higher, so the photon density is 1% higher. The density perturbation by isocurvature results in the sum of surplus densities at each site being zero. If at one location there is 1%

---

*119) In the Big Bang theory it is a physical process of reducing the density anisotropy in the early universe, making the universe and the CMB uniform. In the recombination era, 300,000 years after the Big Bang, photons was cooled, reducing the temperature variation. This effect, together with the acoustic oscillation of matter, the Doppler effect and the gravitational effect on electromagnetic waves, played a role in forming the present large-scale structure of the universe.

more energy in baryons than the average, then there is 1% more energy in photons and 2% less energy in neutrinos. This is the pure isocurvature. The density perturbation by isocurvature produces a series of picks at positions where the angle ratio is 1:3:5... On the other hand, the density perturbation by adiabatic produces picks in the ratio 1:2:3...[180] The observation results demonstrate that the density perturbation in the beginning is adiabatic. As the universe expands, the primordial plasma disintegrates and fades, rapidly increasing the free flight distance of photons and still causing inelastic collisions such as Compton scattering.*[120) These effects suppress anisotropy which disappears at very small angles. The separation of photons and baryons took place slowly over a period of time. It was calculated by the anisotropy probe WMAP to be 115,000 years and the CMB was formed for up to 372,000 years.[181]

After being formed, the CMB was transformed under various influences. This transformation is called the late anisotropy or second anisotropy. When the CMB photons began to travel uninterrupted, most of matter in the universe was composed of hydrogen and helium atoms, most of which were ionized. The CMB photons

---

*120) Discovered in 1923 by Arthur Holly Compton (US physicist, 9.1892 ‑ 3.1962), a phenomenon of photon scattering by charged particles. In this process, the energy of photons decreases and the wavelength increases (the Compton effect). Some reduced energy is transferred to the electrons. Compton won the 1927 Nobel Prize in Physics for discovering this phenomenon. This discovery proved the particle characteristics of electromagnetic waves.

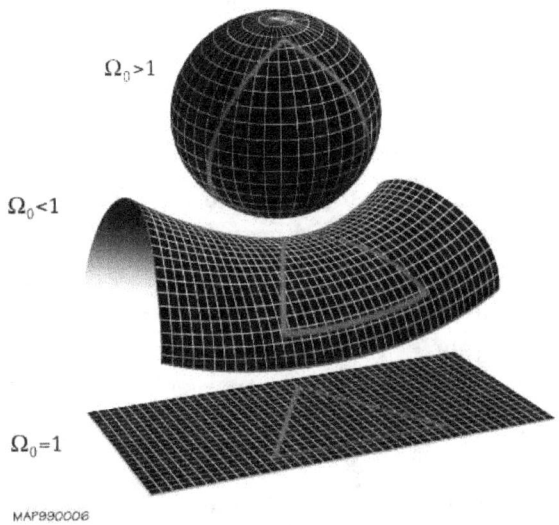

$\Omega_0 > 1$

$\Omega_0 < 1$

$\Omega_0 = 1$

MAP990006

Figure 75. Morphology of the two-dimensional universe with the density factor Ω. The curvature of the universe is determined by Ω (wikipedia.org).

were scattered by charged particles such as free electrons. Today's charged particles in space have a very low density and do not affect the CMB, but in the early universe their density is high, which results in the loss of small anisotropy or the scattering of photons by free electrons. The WMAP spacecraft observed this effect, demonstrating that the universe was in the beginning in an ionized state. However, there are some arguments that the first star, the first supernova, or the radiation of ionized matter from a large black hole disk could be included.

CMB in the new vacuum paradigm

Let us consider the cosmic microwave background, or CMB, in the regime of our vacuum paradigm. First of all, the CMB itself strongly suggests the presence of the solid vacuum. The solid vacuum of the universe contains the CMB energy. Where did the CMB come from? The current story starts from the Big Bang. When matter and light were not separated after the Big Bang (namely, when matter was composed separately of protons and electrons), light could not propagate due to scattering, but when neutral hydrogen was generated, light freely propagated through the universe. It is stated that the CMB currently measured had come from a spherical surface called the surface of the last scatter. We assume here that this spherical surface is the surface of the first supernova. The CMB can then be regarded as attenuated light as it spreads through the solid vacuum from the birth of the first primordial supernova. In this case, the Big Bang did not determine the appearance of the present universe, but the first supernova could be responsible for the present cosmological structure. We imagine that the energy and matter of the primordial supernova gathers again to form small stars, and from some of these stars, smaller supernovae form to give out energy and matter again and repeat this process till now. The anisotropy found in the CMB may reflect this process. Currently, the only interpretation of the CMB anisotropy was made by Wayne Hu, a professor of astrophysics at the University of Chicago. It seems unreasonable to accept his interpretation of the Big Bang theory universally.*121)

Could this anisotropy reflect the structure of the solid vacuum? It is stated that a series of picks in the CMB structure at the angle ratio 1:3:5... originated from the density (baryons, photons, neutrinos) perturbation by isocurvature and 1:2:3... from the density perturbation by adiabatic. If the solid vacuum has a regular lattice structure, as shown in Figure 14, the regularity of the anisotropy may reflect the structure of the solid vacuum. For a steel with a face-centered cubic lattice (fcc) structure, electron diffraction patterns can be obtained by Transmission Electron Microscope (TEM) as exemplarily shown in Figure 76. It is seen that the diffraction is more intense at certain angles when being viewed from the central point.

Abnormal phenomena of the CMB, such as the large scale anisotropy, abnormal alignment, and non-normal (Gaussian) distribution, cannot be explained by the Big Bang theory. Are the phenomena can be interpreted in the regime of our vacuum paradigm?

Finally, if the CMB is present, it is assumed that the CNB (cosmetic neutrino background, or CvB) should be also present. Neutrinos are emitted together with light from stars or from supernovae,[182] but the CNB generated at the beginning of the Big Bang is thought to remain because it rarely reacts with matter. The energy is 10^{-4} to 10^{-6} eV and cannot be observed

---

*121) The Big Bang theory is discussed in detail in Section 3.11.

Figure 76. Electron diffraction pattern of a steel with the fcc structure obtained using a TEM. In the center, the diffraction intensity is strong along certain angles (wikipedia.org).

directly, but there are indirect evidences.[183] If light and neutrinos in the solid vacuum are the source of all matter, it is reasonable to think that the energy density of light and neutrinos was very high before matter was created. Could electrons and positrons form from this light and neutrinos (via vacuum fluctuations or quantum fluctuations, see footnote

106), and then positrons were trapped in the vacuum lattice to be protons, creating matter called baryons? In this case, the CMB and CNB should be leftovers after creating matter in the universe, not from the Big Bang and subsequent expansion of the universe.

4.2. Dark matter, part of the solid vacuum?

Dark matter has been introduced to explain strange phenomena that are not yet defined astronomically, It cannot be observed, as it does not interact with electromagnetic waves. According to the standard model, our universe consists of 4.9% ordinary matter, 26.8% dark matter, and 68.3% dark energy.[184] A fair amount of matter has not been observed yet and adds up to less than 10%.[185] Some astronomers have denied the existence of dark matter in terms of modified theories of General Relativity, but its existence is consistently inferred from galaxy rotation curves, gravitational lensing, and gravitational effects on the CMB.[186] What is dark matter, if any? The most plausible hypothesis is that dark matter is composed of particles that respond to gravity and weak force.[187] Some phenomena cannot be explained without dark matter,[*122)] but the particles in the hypothesis have not yet been found.

----

*122) In particular, the explanation of the anisotropy in the CMB.

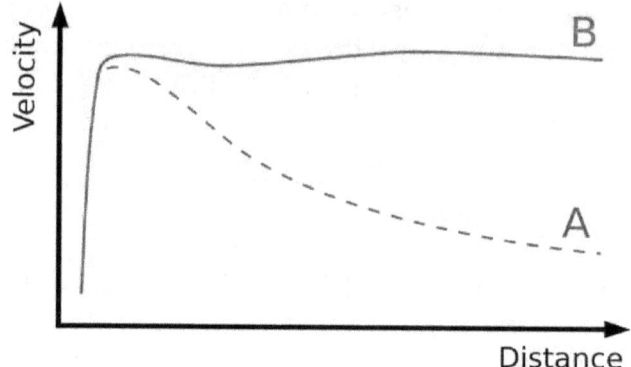

그림 77. Velocity curve in spiral galaxies: calculation (A), observation (B) (wikipedia.org).

Why dark matter should exist

Stars in a galaxy follow the so-called virial theorem.*123) The rotational velocities of stars of a galaxy can be calculated directly from the Doppler effect of light. The average velocity $v(r)$ of a star, away from the center of a galaxy cluster by the distance r, has the virial relationship with the total mass $M(r)$ contained in the sphere of radius r. In 1933, Zwicky

---

*123) In general dynamics, the average kinetic energy and average potential energy are proportional to each other. The virial theorem is used to calculate the average total kinetic energy of a system that is impossible to solve analytically. If a system is in a thermal equilibrium, the average kinetic energy is statistically related to the temperature of the system. The theorem can also be applied to systems that are not in thermodynamic equilibriums regardless of the definition of temperature.

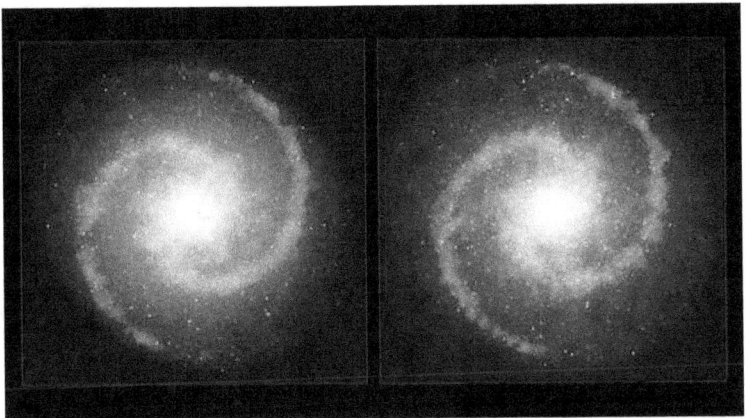

Figure 78. (left) Virtual current galaxy shows high rotational speeds at the outer skirt containing dark matter (looks blurry). (right) An imaginary galaxy billion years ago shows relatively slow rotational speeds on the outer skirt. Dark matter may not exist if the mass of the black hole supposedly at the center of the galaxy is very small (wikipedia.org).

*124) first applied the virial theorem to the Coma Cluster*125) and calculated the mass due to gravity in this cluster.[188] The result gave 400 times more than the mass of the galaxy cluster, and the unknown invisible matter making up this difference was called dark matter.

The mass of the illuminant of a spiral galaxy

---

*124) Fritz Zwicky (2.1898 – 2.1974) Swiss astrophysicist.
*125) Also known as Abel 1656, this galaxy cluster is a giant cluster of over 1,000 galaxies. This galaxy cluster is named after the constellation because it is in the Coma Cluster.

Figure 79. Light is deflected by gravitational lensing of a giant structure. The white arrow is the deflection of light due to gravity, and the orange one is the trajectory of light seen by an Earth observer (wikipedia.org).

decreases on going away from the center. According to the observation, $v(r)$ decreases more slowly than the mass distribution, being confirmed by the Doppler effect of light. This may mean that there is a huge mass in the periphery of the galaxy that cannot be observed by electromagnetic waves. If the illuminant contributes most of the galaxy's total mass, the rotational speed at the outer edge would be reduced according to Kepler's second law. However, the speed does not decrease and is almost constant as shown in Figure 77.[189] If Kepler's law is correct, there should be a significant amount of non-luminous matter. One can determine the mass distribution of a galaxy from the observed rotational speeds of stars in the galaxy and calculate the velocity distribution using the mass distribution of the stars and the virial theorem. The velocity distribution predicted from the mass distribution and the actual velocity distribution are different. This is a reason why dark matter must exist.

Gravitational lensing is another reason for the presence of dark matter. Gravitational lensing is an effect due to which light is bent by massive objects such as galaxy clusters between the distant light source and the observer (see Figure 79). General Relativity can predict this kind of deflection.*[126] In 1937, Zwicky proposed gravitational lensing by galaxies,[190] but only in 1979 was first identified by a pair of quasars, "Twin QSO" SBS 0957 + 561.[191] Figure 80 shows a strong gravitational lensing effect observed in the Abell 1689*[127] galaxy cluster by the Hubble Space Telescope.[192] By measuring the distortion, the mass of the interstellar cluster can be obtained[193] and the density of dark matter can be estimated from the mass-to-light ratio.[194]

Dark matter does not interact with electromagnetic waves, but leaves traces of the gravitational influence on the CMB. The CMB is like a perfect black-body

---

*126) As demonstrated in section 2.7, the same prediction can be made in the regime of our vacuum paradigm.
*127) Giant galaxy cluster in Virgo. About 12 billion light-years from Earth. It consists of more than 160,000 galaxies.

Figure 80. A gravitational lensing effect observed in the Abel 1689 galaxy cluster by the Hubble Space Telescope (wikipedia.org).

radiation, but has a slight anisotropy in the temperature distribution. This can be resolved into an energy density spectrum as shown in Figure 74. The first pick indicates the anisotropy due to the effect of the final scattering surface. The second is the effect of the interaction of the CMB with hot gases or gravity, which occurs between the last scattering surface and the observer.[195] The third one is stated to be related to the density of dark matter.[196]

What is dark matter in the solid vacuum?

There are additional phenomena which can only be explained only by dark matter such as the structure formation of the universe, but will be omitted in this book. The three reasons mentioned above are summarized as follows. The first is that, as shown in Figure 77, the velocities of stars in a spiral galaxy do not decrease and remain constant on going away from the center. Dark matter well explains this abnormal velocity distribution. The premise here is that the center of the galaxy has a very massive black hole. In section 3.7 where we dealt with black holes, we argued that such black holes cannot exist and, if any, they are just "black" holes free of energy, made up with the cold solid vacuum. In General Relativity, the gravitational attraction of black holes is so great that even light is absorbed. In our vacuum paradigm, a black hole is the remnant after a neutron star consumes its energy. Namely a black is just what remains after the complete incineration of a neutron star and eventually dies out. It appears to be a "black" hole because the remaining energy is almost exhausted. Because a black hole is black, they absorbs light well like an ideal black-body, and because it is just an empty space without matter, it will swallow matter in the neighborhood. Its surrounding may shine brightly due to the incoming materials as shown in Figure 64. If the mass of the black hole at the center of a galaxy is very low, the speeds of the stars will not simply decrease as the distance increases from the center. Since the velocity is proportional to the square root of the mass contained in the radius of orbit, as in Eq. (2.15), if the velocity is very low at the center and the mass increases from a certain distance from the center, the velocity will not decrease. The constant velocity means that the mass is proportional to the distance from the center according to equation (2.15). As is consistent with our inference that at the center of the galaxy the mass should be the lowest and will increase with distance. As Figure 79 shows, the galaxy from billions of years ago had a higher mass in the center, but the mass in the center of the current galaxy is lowered so that, it is reasonable to assume, the velocities of stars farther from the center have increased.

Another reason for the abnormal rotational velocity distribution of spiral galaxies can be explained from the fact that the cross section of our galaxy is very flat, as shown in Figure 44. The distortion of the solid vacuum induced by a rest object is three-dimensionally point-symmetric, but if it rotates as the case with a spiral galaxy and the mass is concentrated on the plane of rotation, the distortion develops in a two-dimensionally point-symmetric manner. This will increase the distortion of the solid vacuum in the plane of rotation and increase the influence among the stars lying on this plane. The larger the flattening of the spiral galaxy, the greater its influence and ultimately the constituent stars will move as one body. Going away from the center, the velocity of the stars increases in proportion to its distance. Thus, the B curve in Figure 77 reflects the combination of the decrease in velocity by Kepler's law (A curve) and the increase due to the two-dimensional distortion of the solid vacuum of a flat spiral galaxy. We have already noticed such a influence of the distortion of the solid vacuum in the precession of Mercury in section 2.8.

The second suggestion for the existence of dark matter is gravitational lensing, as shown in Figures 79 and 80. By measuring the distortion of the image of the rear galaxy due to gravitational lensing, the mass of the intermediate cluster is obtained to prove the existence of dark matter from the mass-to-light ratio. If there is mass, there is light, and if the mass corresponding to the luminous intensity differs from the mass calculated from the image distortion, the difference should be back up by the mass of dark matter in the regime of General Relativity. The mass due to the distortion can be obtained not only from General Relativity, but in the regime of our vacuum paradigm as discussed in Section 2.7. Two questions are raised here. Do all the mass emit light? We know that the Sun only shines brightly itself. In fact, non-luminous masses are estimated to be less than 10% to the total mass of the universe. So it is claimed that dark matter must exist. The other question is whether a gravitational lens is concave or convex. As shown in Figure 81, the incident light beams are magnified over a long distance regardless of the type of lens. In the case of the convex lens, the image is seen reduced just after being conducted though. We

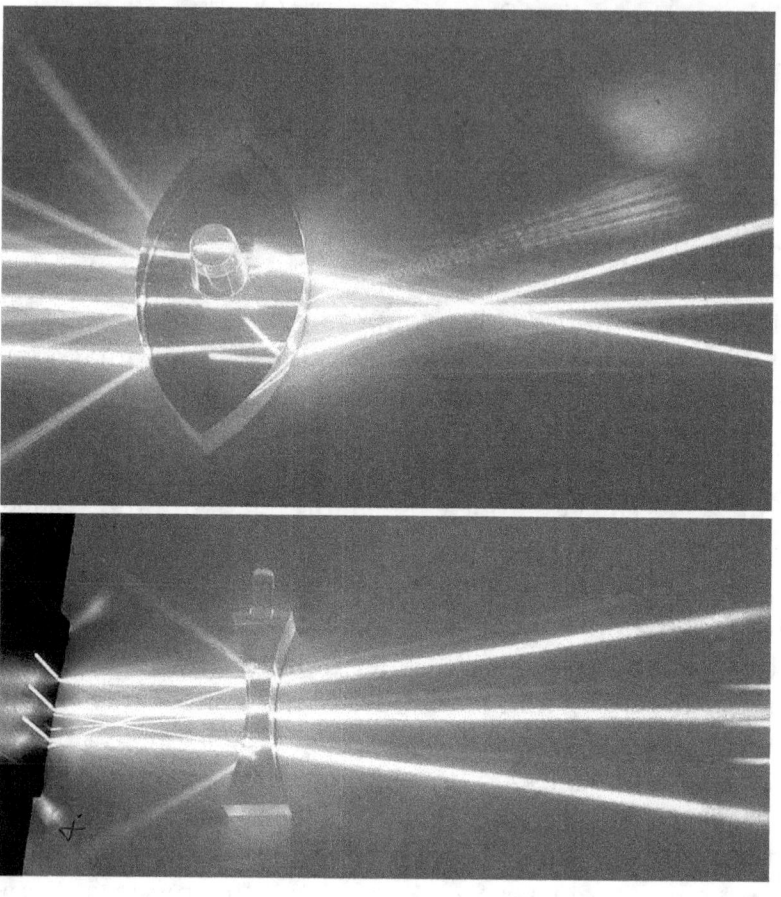

Figure 81. Deflection of light by convex (top) and concave (bottom) lenses (wikipedia.org).

can say that the origin of gravitational lensing is not because of dark matter, but a black hole with no mass in the cluster.

People look for the third proof of dark matter in the CMB. It is claimed that dark matter makes an influence on the anisotropy of the CMB, as shown in Figure 74. This is related to the Big Bang model, in particular the Λ-CDM model, the standard model of the current universe. But just as the cosmological constant was introduced to make up the arguments on General Relativity, dark matter was introduced to avoid the controversy of the Big Bang model. The introduction of something that doesn't really exist can be a conventionalist stratagem to hide and protect the contradictions of the current theory.[197]

4.3. Dark energy – energy contained in the solid vacuum?

In modern astrophysics and astronomy, dark energy is a kind of unknown energy that is thought to permeate all over the universe and causes the accelerated expansion of the universe.[198] If the Λ-CDM model of the universe is accurate, it accounts for 68.3% of the whole energy in the universe. The dark energy density is very low, $\sim 7 \times 10^{-30} g/cm^3$, but the absolute amount is large because it is spread throughout the whole universe.[199] The candidates for dark energy are the cosmological constant Λ that are evenly distributed throughout the universe, or scalar like quintessence, or any dynamic quantity whose energy density changes in time and space.[200] Energy that does not change in space and has a constant value is included in the cosmological constant and is identified as vacuum energy.[201]

Origin and evidence of dark energy

Λ was introduced and included in Einstein's equations of (1.9) to ensure that the universe does not contract due to gravity. However, this static universe has been found to be unstable because local anisotropy eventually causes an expansion or a contraction.[202] When the universe expands slightly, it expands because it releases vacuum energy, and when contracted slightly, it contracts further. This is inevitable because of the uneven distribution of matter in the universe. Furthermore, Einstein admitted his failure about his assumption of the static universe in 1929 when Hubble observed evidences for the expanding universe.*[128) Since then, Λ was considered unrelated to the present universe for a while. However, in 1998 and 1999, when the redshift of distant supernovae indicating the accelerated expansion of the universe were observed, a constant in the cosmological theory was required, and

---

*128) Gamow, George (1970) My World Line: An Informal Autobiography. p. 44: "After a lot of time, when discussing the issue of the cosmological constant with Einstein, he confessed that the introduction of the cosmological constant was the biggest failure in my life." The "cosmological constant" here is a constant in the equation of General Relativity introduced by Einstein to ensure that the universe neither expands nor contracts. Ironically, this is the theoretical prediction of Hubble's first observation of the expansion of the universe.

imaginary dark energy represented by this constant was born.[203]

In General Relativity, the extent to which the universe expands can be estimated from the curvature and the equation of state of the universe.[*129)] Currently, the Λ-CDM model is regarded as the "standard model of cosmology" because it best matches today's cosmological observations. However, it is not enough to confirm the existence of dark energy only through observations of the universe far away from us. Regardless of its reality, dark energy must have a strong negative pressure (repulsive force), such as the radiation pressure, to explain the accelerated expansion of the universe.[204] The existence of dark energy is confirmed indirectly based on the following three things:

- Relationship between distance and redshift: an evidence that the universe is acceleratedly expanding.
- The third energy, not matter nor dark matter, is required to explain the theoretically flat universe.[*130)]
- Large-scale fluctuation patterns of the mass density measured in space.

The reason for the existence of dark energy

Regardless of its form, dark energy is required to

---

*129) It describes the relationship among temperature, pressure, and density of matter, energy and vacuum energy for any space in the universe.
*130) There is no evidence that the current universe is curved.

match with the geometry of space observed to date. It is seen from the CMB anisotropy that the universe is nearly flat (three dimensionally flat). The mass/energy density of the universe should be equal to the critical density, as shown in Figure 75. The critical density is the density of the flat structure of the universe in the Friedmann–Lemaître universe.*131) At present, the critical density is calculated to be about 5 hydrogen atoms per square meter, but the actual mass density is estimated to be 0.2-0.25.205 Therefore, if the universe is flat, it needs to be filled with something other than matter, and it should be dark energy, since the total of matter plus dark matter is only about 30% of the estimation.206 Recent observations of supernovae indicate that the universe is made of 71.3% dark energy and the rest 27.4% is the sum of dark matter and matter.207 The CMB data analysis from the European CMB probe Planck spacecraft estimated the amount of dark energy 68.3%, dark matter 26.8%, and only 4.9% for ordinary matter.208

The simplest explanation for dark energy is the intrinsic fundamental energy of the vacuum, the cosmological constant Λ. By mass-energy equivalence, this energy influences gravity. The universe acceleratedly expands because Λ exerts negative pressure as much as this energy. When a space expands thermodynamically, its energy decreases (the

---

*131) The Friedmann-Lemaître universe is a uniform, isotropically expanding universe assumed by the Friedmann-Lemaître -Robertson-Walker (FLRW) metric, an exact solution of Einstein's field equations. This evolved into the Λ-CDM model.

vessel must work outside). However, the pressure becomes negative because the vacuum energy increases by volume expansion. This cosmological constant is a key factor in the Λ-CDM model. However, quantum field theory also estimates the vacuum energy of more than 100 powers,[209] a huge difference from the cosmological constant. This difference is called the vacuum catastrophe. Some supersymmetric theories require that the cosmological constant be exactly zero.[210] Skepticism on dark energy is strictly present and denies its existence. It is also argued that dark energy is a mirage that does not exist. Analyzing the observations more carefully based on the current theories may indicate that dark energy is just a measurement error. For example, if we are in a space with a density less than the average, the observed cosmic expansion is a mistake in the change in time or in the acceleration.[211] Using the equivalence principle that gravitational mass and inertial mass are the same, the space appears to expand more rapidly in the void around a star cluster. No matter how insignificant this effect will be, if billions of years accumulates, it will make the universe appear expand and make us seem to be in the Hubble's bubble.[212] Another view is that the accelerated expansion of the universe is an illusion formed by our relative motion to the universe.[213]

To summarize the above mentioned, dark energy was introduced to explain the hypothesis that the observed redshift from distant supernovae is due to the accelerated expansion of the universe. It is represented by the cosmological constant Λ, a constant introduced into Einstein's field equations of General Relativity. Our universe will not contract by gravity in the presence of dark energy. One of the exact solutions of the field equations, the FLRW metric assumes our universe to be a uniform and isotropically expanding one. The anisotropy of the CMB is claimed to indicate that the universe is flat, which in turn predicts 70% dark energy to the total matter/energy of the universe. Dark energy is claimed to trigger an accelerated expansion in the regime of the FLRW metric, as is confirmed by the redshift of the distant universe.

Dark Energy in the solid vacuum

Gravity whose origin is unidentified is the key of General Relativity, represented by Einstein's field equations.*132) In our vacuum paradigm matter currently observed in space is regarded as energy stored in the vacuum lattice, which in turn distorts and expands the solid vacuum. The distorted solid vacuum in the presence of matter yields an attractive force, gravity. It is similar to the concept of the distortion of spacetime in General Relativity. Due this apparent attractive force, matter aggregates to form stars. However, as the energy of stars is squeezed out exerted by the surrounding solid vacuum, they become smaller in size or explode to supernovae,

---

*132) Einstein said that gravity is not a force but a phenomenon that comes from the spacetime distortion. See section 1.4.

leaving only the cores to be neutron stars. Starlight is part of the energy released by this process. The distorted solid vacuum induces an attractive force (it is called gravity) for matter to aggregate, but also exerts pressure on matter so that the energy of aggregated matter is squeezed out again into the surrounding cold solid vacuum. In the regime of our vacuum paradigm, there is no reason for the universe to contract due to gravity and the cosmological constant is not required to prevent this contraction. The contraction or expansion of the universe would not be relevant in our vacuum paradigm.

If dark energy is exist, how is it related to the solid vacuum, and how are the phenomena that confirm its existence related to the behavior of the solid vacuum? The essence of our vacuum paradigm is the solid vacuum. The existence of this medium cannot be directly confirmed. This is because the solid vacuum is the fundamental and basic structure free of energy (it is a platform for mass and energy). What we can see is what happens when energy is added to the cold solid vacuum. In that sense, dark energy should be a kind of energy stored in it. The size of the universe is unknown, but if a tiny energy, such as the CMB, is spread throughout the universe, it will be huge. One of the sources of dark energy may be neutrinos. Neutrinos are something with energy, but does not react with matter, a similar property to dark energy. Neutrinos are released continuously with light from stars including supernovae, from which a huge amount of neutrinos is emitted. But we still do not know how neutrinos are stored in the solid vacuum. In the nuclear fusion process shown in Figure 43, neutrinos are released as electrons and protons fuse to neutrons. Even the remained energy of these neutrons is exhausted, we have only the cold solid vacuum not distorted and thus with no energy. The emitted light energy will remain as the CMB we observe, and neutrinos will be dark energy that is currently unobservable. This is the cosmic neutrino background, the CNB, as compared to the CMB. Since neutrinos do not generate electric and magnetic fields by vibrating the vacuum lattice in a shear mode, they rarely react with matter. When neutrinos accumulate (namely, when energy accumulates) in the solid vacuum of the universe, the entire universe can be thought to be (thermally) expanding. To date, the speed of neutrinos is known to be the same as that of light. We hypothesized in Section 2.2 that neutrinos are the pressure waves in the solid vacuum and that electromagnetic waves are the shear waves. If neutrinos are indeed the pressure waves, the bulk modulus should be negative, as given in Eq. (2.5). Thermodynamically, when a space expands, energy decreases in it because it works against the outside. Therefore, the negative bulk modulus is thermodynamically impossible. However, because dark energy, represented by the cosmological constant, increases by volume expansion, the pressure becomes negative. In this sense, dark energy appears to be related to the negative bulk modulus of the solid vacuum.

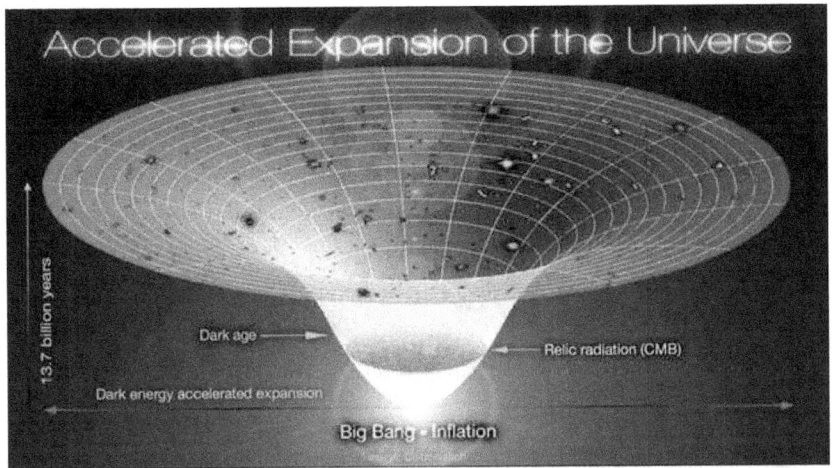

Figure 82. The accelerated expansion of the universe in the Λ-CDM model (wikipedia.org).

4.4. Instead of the Big Bang theory

The Big Bang theory is now accepted as well established one, explaining the origin of the universe. According to the theory, a big explosion occurred at one point where the universe was concentrated 13.7 billion years ago and continued to expand.[214] This theory is based on Hubble's observations of the universe, indicating that distant galaxies continue to move away from us, as inferred from the observed redshift. Reverting the clock of the universe, which is expanding at the current rate, leads to a conclusion that the universe would have be concentrated at one point about 13.7 billion years ago.

The Big Bang theory - Λ-CDM

Most theories on the universe start with the assumption that it is uniform no matter where we ar e.[215] Currently, the most prominent Λ-CDM model is theorized by introducing the cosmological constant Λ as dark energy and cold dark matter (CDM) into the Big Bang hypothesis. This model appeared in the late 1990s and is based on General Relativity. This model well explains the following phenomena, as described in the early sections of Chapter 4.

- The CMB anisotropic structure
- The large scale structure of galaxy distribution
- Large amounts of hydrogen, helium and lithium throughout the universe
- The accelerated expansion of the universe, as observed in distant galaxies and supernovae

The period when stars, galaxies, and nebulae were formed from primordial matter after the Big Bang is called the structure formation epoch. According to the Friedmann solution, the FLRW metric, the universe was uniform prior to this epoch.[216] Matter is influenced by light, the main energy of the early universe, but because it is irrelevant to density fluctuations, no structure could be formed.[217] The uniformity was then broken by dark matter, allowing stars, galaxies, and other large-scale structures to form. If there were only matter in the universe, there would be no time for nebulae or galaxies to form. As dark matter does not interact with light, density fluctuations led to a non-uniform density distribution of matter. Thereafter, matter gathered together under the influence of gravity, so that the present universe was formed.[218] This process shall be inferred from the currently observed anisotropic structure of the CMB. This model, according to which the universe is acceleratedly expanding, is based on time delay or time dilation resulted from the light decay of supernovae and on the redshift from distant galaxies. Both phenomena are confirmed by the Doppler effect. In order for the universe to acceleratedly expand, we need unknown energy that acts as a force (negative pressure) to offset the gravitational attraction. This shall be the cosmological constant Λ.[219] CDM and Λ are required to explain the present structure and the future destiny of the universe. So the standard model of the current universe is the Λ-CDM model. It is claimed.

What is missing in the Big Bang

The Big Bang theory that the universe was born with a huge explosion at one point where the energy of the whole universe was concentrated is hard to believe in common sense. However, the presence of large amounts of hydrogen, helium and lithium in the current universe is allegedly due to primordial or Big Bang nucleosynthesis that occurred in the early stages of the Big Bang. Nuclear fusion, which took place between 10 seconds and 20 minutes after the Big

Bang, synthesized light elements other than hydrogen isotopes.[220] At present, 75% hydrogen and 25% helium in the universe roughly agree with the predictions made by the Big Bang nucleosynthesis. The Big Bang theory predicts the amount of lithium, but the observed amount is only one-third of the prediction. Some scientists assume that the actual amount of lithium is small compared to the theoretical estimate because there is a high probability of lithium to convert to other elements.[221] This hypothesis has not been proven experimentally due to its difficulty. In a recent study, Kawabata's team at Kyoto University of Japan found using unstable helium that beryllium is unlikely to turn into helium.[222] This result contradicts the current explanation regarding to the gap between the theoretical and measured amount of lithium. The Big Bang theory is being challenged. It should be scrapped or revised.

Until now, dark matter has not been found, but exists only in theory. The density of dark energy is too small to be observed. This model accounts for the observations of the large-scale cosmological structure, but it predicts too much dwarf galaxies*133) below the galaxy scale and too much dark matter at the very center of the galaxy.[223] This problem cannot be solved by computer simulation. It is still unclear whether this is a problem of computer simulation itself, misunderstanding of the characteristics of dark matter, or a serious defect in the model. Karl Popper,

---

*133) It is a small galaxy composed of up to billions of stars. For reference, our galaxy consists of 200 to 400 billion stars.

Figure 83. Karl Popper (6.1902 – 9.1994). He was an Austrian-born British philosopher and professor. He is regarded as one of the 20th century's greatest philosophers of science.

one of the great 20th-century philosophers of science, criticized that there is no room for contradiction in the Λ-CDM model, because it was organized based on conventionalist stratagems.[224]

New cosmology in the regime of new paradigm

In the regime of our vacuum paradigm, we can assume that the solid vacuum as well as matter were born at the very beginning of the Big Bang, if we accept the Big Bang as real. Alternatively, it is assumed that only matter was generated in the already existing solid vacuum at the time of the Big

Bang. In the latter case the expansion of the universe does not hold. Because the "clean" universe made of the cold solid vacuum already was there. The current Big Bang theory represented by the Λ-CDM model has unidentified or contradictory problems such as dark matter, dark energy and the amount of lithium in the universe. This theory is based on General Relativity. As the spacetime distortion of General Relativity is similar to the distortion of the solid vacuum, there is no significant difference in the interpretation of the cosmological phenomena. However, General Relativity does not tell the origin of gravity. It does not explain how inertial motions of matter is directly related to gravity. In regime of our vacuum paradigm, gravitational mass is anisotropic depending on the direction of the motion of matter in the solid vacuum. Gravity is not an inherent fundamental force due to the presence of mass, but an apparent one originated from the distorted solid vacuum and matter wave. Massive objects feels pressure from the surrounding solid vacuum. Cosmic phenomena such as supernovae and black holes were understood in terms of the pressure exerted by the distorted solid vacuum consistently. How was the universe created in the regime of our vacuum paradigm, if not for the Big Bang theory like the Λ-CDM model?

We propose a theory of the recurring universe instead of the Big Bang theory. Matter is interchangeable via mass-energy equivalence. We should say that there is no difference between them. They are all just the vibrations of the solid vacuum lattice. If we distinguish substances in terms of vibration, there are things that spread through the solid vacuum without hanging out in one place, such as light or neutrinos, and there are things bound to or staying around the vacuum lattice, such as electrons, protons and neutrons. It is (rest) mass which is bound to the vacuum lattice.

First, let's say that light and neutrinos are in their fullest amount in the universe made of the cold solid vacuum. The solid vacuum is fluctuating evenly throughout the whole universe. Due to this fluctuation the universe is in its greatest expansion, with electrons and positrons repeatedly being produced and destroyed locally. Some positrons generated are trapped in the vacuum lattice by absorbing neutrinos. Electrons and protons are generated in this way, and they are attracted to each other in the distorted solid vacuum, yielding hydrogen atoms. As light and neutrinos are converted to mass, the universe cools down and its size decreases. More and more hydrogen atoms are generated and the universe is getting cooler. Several hydrogen atoms combine together to reduce the distortion of the solid vacuum and eventually form a big sphere composed almost of hydrogen atoms. One very big primordial superstar is born. This superstar is more massive than all the stars of the present universe. This only one star in the universe explodes to be an enormous super-supernova, through the same process as currently observed supernova explosions. A great amount of new stars are born and thus smaller supernova explosions continue. The remnants makes up all the stars, galaxies, and so on, with considerable energy being released into the universe in the form of light and neutrinos. The universe is warming up again. The matter scattered throughout the universe is consumed by releasing both light and neutrinos via the life-cycle process of stars in the influence of the distorted solid vacuum, and the universe becomes a place filled only with light and neutrinos. When the energy of cosmic light and neutrinos exceeds the critical point, electrons and protons are regenerated in the solid vacuum, and the energy of light and neutrinos are converted into mass. Thus it completes a cycle of the recurring universe. This process may be repeated again and again. Another scenario is that an equilibrium may be reached at some point, a dynamic equilibrium where mass is created from light and neutrinos on one side and mass returns to light and neutrinos on the other. Forever...

V. The universe in the solid vacuum is simple and clear

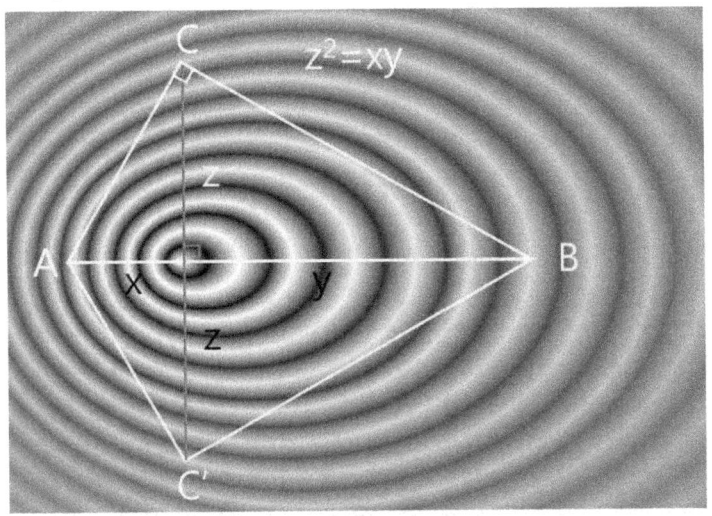

Mass is energy and this energy is the vibrational energy of the solid vacuum. This vibration emits point symmetrical matter wave, and the greater the mass, the higher the frequency. When the center of vibration moves, the symmetry of matter wave is broken and stored as the distortion energy of the solid vacuum. Objects always move towards the high distortion side of the solid vacuum. The anisotropic distortion remains constant as a result of this movement, and the movement continues. It is inertia. Point-symmetric distortion around an object induces asymmetrical distortion around other nearby objects and thereby attracts each other. The closer the objects are, the greater the attraction force is, so that in the interior of a star nuclear fusion occurs continually under very high pressure and its energy is released slowly or explosively as supernovae in the form of light or neutrinos.

We have reinterpreted cosmological phenomena in this book, assuming that the vacuum has no energy and is made of a very dense medium (we called it the cold solid vacuum). When energy is added to the medium constituting the vacuum, the solid vacuum deforms and distorts, yielding stress/strain not only in the solid vacuum but in ordinary matter. The strained or deformed state of the solid vacuum is the origin of all the phenomena in nature including gravity. The solid vacuum does not change itself, but its energy state due to the vibration of the lattice changes, and matter or energy transfer is the movement of the lattice vibration energy through the solid vacuum. From this premise, we could identify the origin of gravity and consistently interpret other cosmological phenomena.

Assuming that the vacuum is made of a very dense medium (the density is maximized when there is no energy), and matter such as atoms, planets, stars, etc., has energy and thus push the solid vacuum as much as the energy regardless of the size, the surrounding solid vacuum will be distorted and the intensity of the distortion will decreases on going away from the center of matter. This distortion of the solid vacuum is similar to the concept of the spacetime distortion by mass and energy in the regime of General Relativity. Gravity is an apparent attractive force arising in the distorted solid vacuum, as the energy of matter pushes the surrounding solid vacuum.

The distorted solid vacuum determines the behavior of matter. In Newtonian mechanics the law of inertia states that an object moves in the vacuum at a constant velocity, if not interfered by any other forces. The movement of the object inevitably leads to an asymmetric distortion of the solid vacuum. As can be confirmed by the Doppler effect of light, the solid vacuum is compressed in the forward direction and relaxed in the opposite direction of the moving object, giving rise to a difference in the distortion of the solid vacuum between both ends of the moving object. This difference is the driving force for an inertial movement and also the origin of gravity. When two objects are located nearby in space, one object is influenced by the distortion of the solid vacuum developed by the other. Since the strength of the distortion decreases with distance, the object feels an anisotropic distortion of the solid vacuum around it along the line connecting the centers of the two separated masses. One object will move towards the other because of this difference, according to the law of inertia. If no other force is applied, the movement increases in speed, as the difference in the distortion of the solid vacuum will increase as one object approach the other. Namely, the movement is accelerated, and the object feels an attractive force inversely proportional to the square of the distance. This is the very origin of gravity.

Why do objects move upward regarding to the distortion of the solid vacuum? This is because the object vibrates constantly as much as the energy in the solid vacuum and moves in interaction with the surrounding vacuum lattice. This vibration is matter wave. The wavelength of matter wave is the modulated lattice spacing of the solid vacuum at the interface between matter and the solid vacuum. The greater the mass, the greater the speed of movement of matter, the narrower this lattice spacing. Thus, the wavelength of matter wave is inversely proportional to the product of mass and velocity of matter, namely momentum. When the solid vacuum is compressed, that is, the wavelength of matter wave is shortened, the distance between the two adjacent vacuum lattice is narrowed, so that the probability of movement due to vibration increases. This is the law of inertia where objects move in one direction and also the very origin of gravity.

The distortion formed in the solid vacuum yields attraction between separated objects, due to which they are attracted to agglomerate. As matter aggregates, the compressive stress inside matter increases, creating infinite stress at the central point in a spherical object. If the interior of matter is at a very high pressure, energy will squeeze out and the object will shrink. Nuclear fusion turns protons and electrons into neutrons and releases energy to warm up surrounding matter in the form of electromagnetic waves and neutrinos. Stars including the Sun is mainly composed of hydrogen. Their energy is radiated as hydrogen is fused into helium. Earth-like planets are made up of heavy elements such as iron, and continue to release energy, although their fusion reactions are not active. Energy is released in the form of earthquakes or volcanic activities on Earth.

When the internal pressure is very high in massive stars, the fusion reaction is very active, but when the energy release is suppressed, it explodes without enduring the accumulated energy. The explosion of a massive star is seen in the form of supernovae. A supernova releases heavy elements formed inside the progenitor star into space, leading to the formation of planets like Earth made mostly of heavier elements than hydrogen and helium. There are unstable atoms in Earth which originated from the fusion process in massive stars, and when these atoms are exposed in space, the nuclear fission reaction occurs, the reverse reaction of nuclear fusion. We can understand why we have nuclear materials on Earth. Both the fusion and fission reactions are processes of releasing energy, but their origins are the opposite.

A neutron star is one of the remains of a supernova, the explosion of a massive star, produced inside the progenitor star before explosion. It is very dense and is regarded as a very large element mostly composed of neutrons that have lost some or all of the energy of protons and electrons. A neutron star can release its compressive energy developed before supernova explosion because the surrounding materials are scattered away along with the explosion. Some high-energy neutron stars spin very fast and emit pulsed strong electromagnetic waves into space. As a neutron star releases energy, it slowly cools down, turning into a black hole and completely extinguishes. A black hole is a state in which almost all the energy of the progenitor star is lost. When even the remained faint energy is consumed, the black hole will eventually be a part of the cold solid vacuum. A black hole is not a monster that sucks everything including light, but it is just a "black" part of the cold solid vacuum free of energy.

Today's universe is allegedly full of dark matter and dark energy that we do not know. It is claimed, because of dark matter, the velocity of stars does not decrease as they move away from the center of a spiral galaxy, and the distant galaxies move faster away from us because of dark energy. These unidentified dark matter and energy should have determined the morphology of the current universe after the big explosion at one point 13.7 billion years ago, and shall make it expand further. Is that so? In the new universe full of the solid vacuum, dark matter is just another description of the two-dimensional distortion of the solid vacuum, which is concentrated on the plane of rotation of a spiral galaxy. The distortion of the solid vacuum on the plane of rotation increases the interaction among the stars on that plane and suppresses their slowing down with an increase in the distance from the galactic center. It is the same principle as that of the Mercury's precession. The effect of the two-dimensional distortion is more pronounced when there is a black hole in the center of a galaxy, which is just a part of the solid vacuum deprived of mass and energy in the regime of our vacuum paradigm.

Dark energy is the cosmological constant included in the equations that describes General Relativity.

According to a solution of the equations, the universe is predicted to expand acceleratedly, being supported by the redshift from distant galaxies. In the universe made of the solid vacuum, there is no such expansion. What appears to be a distant galaxy moving away is that the galaxy is actually moving acceleratedly away from us, or the amount of energy emitted decreases, resulting in longer wavelengths of light, or both. There is no dark energy leading to the accelerated expansion of the universe in the solid vacuum. The universe is not expanding, but matter and energy are spreading throughout the solid vacuum.

Our universe is full of microwaves. The cosmic microwave background is a very long-wavelength electromagnetic radiation that mildly shakes the solid vacuum. By tracking the source of this radiation, we come across a surface and time of the first radiation. It is the explosion of the only one primordial giant star in the universe, the first super supernova. Debris of this very big supernova was spreading out into space and formed the shape of the current universe. The primordial universe was filled with light and neutrinos in the solid vacuum. Electrons and positrons were generated from light and neutrinos, and positrons were trapped in the vacuum lattice, producing protons. Matter was born. Electrons were attracted to protons to form hydrogen atoms, which united into one giant star. The present universe is just the remnant after this progenitor's explosion.

References

1 "Neutrinos sent from CERN to Gran Sasso respect the cosmic speed limit, experiments confirm" (Press release). CERN. 8 June 2012. Retrieved 2 April 2015.

2 Superforce: the Search for a grand Unified Theory of Nature. New York: Simon and Schuster. p. 104

3 T. Young, (1804), "Bakerian Lecture: Experiments and calculations relative to physical optics". Philosophical Transactions of the Royal Society. 94: 1-16.

4 - Huygens (1690), translated by Silvanus P. Thompson (1912), Treatise on Light, London: Macmillan, 1912; Project Gutenberg edition, 2005; Errata, 2016.
   - D.J. Leiter, S. Leiter (1 January 2009), A to Z of Physicists. Infobase Publishing. p. 108. ISBN 978-1-4381-0922-0. Retrieved 11 May 2013.

5 A.A. Michelson, E.W. Morley (1887), "On the Relative Motion of the Earth and the Luminiferous Ether". American Journal of Science. 34: 333-345.

6 - C. Eisele, A,Y. Nevsky, S Schillerv (2009), "Laboratory Test of the Isotropy of Light Propagation at the 10-17 level" . Physical Review Letters. 103 (9): 090401.
   - S. Herrmann, A. Senger, K. Möhle, M. Nagel, E.V, Kovalchuk, A. Peters (2009), "Rotating optical cavity experiment testing Lorentz invariance at the 10-17 level". Physical Review D. 80 (100): 105011.

7 J. van Dongen (2009), On the role of Michelson-Morley experiment: Einstein in Chicago Archive for history of exact sciences 63: 655-663

8 - G. Sagnac (1913), "L'éther lumineux démontré par l'effet du vent relatif d'éther dans un interféromètre en rotation

uniforme" [The demonstration of the luminiferous aether by an interferometer in uniform rotation], Comptes Rendus, 157: 708-710
- G. Sagnac (1913), "Sur la preuve de la réalité de l'éther lumineux par l'expérience de l'interférographe tournant" [On the proof of the reality of the luminiferous aether by the experiment with a rotating interferometer], Comptes Rendus, 157: 1410-1413

9 M. von Laue (1911), "Über einen Versuch zur Optik der bewegten Körper". Münchener Sitzungsberichte: 405-412. English translation: On an Experiment on the Optics of Moving Bodies.

10 M. Dalarsson, N. Dalarsson (2015), Tensors, Relativity, and Cosmology (2nd ed.). Academic Press. p. 106-108.

11 - G.F FitzGerald (1889), "The Ether and the Earth's Atmosphere", Science, 13 (328): 390,
   - H.A. Lorentz (1892), "The Relative Motion of the Earth and the Aether", Zittingsverlag Akad. V. Wet., 1: 74-79

12 A. Pais (1982), Subtle is the Lord: The Science and the Life of Albert Einstein, New York: Oxford University Press, ISBN 0-19-520438-7

13 S. Adams (1997), Relativity: An introduction to spacetime physics. CRC Press. p. 54. ISBN 0-7484-0621-2.

14 A. Einstein (1905), "Zur Elektrodynamik bewegter Körrper". Annalen der Physik. 322 (10): 891-921.

15 N. Ashby (2003), "Relativity in the Global Positioning System" (PDF). Living Reviews in Relativity. 6: 16.

16 J. Forshaw, G. Smith (2014), Dynamics and Relativity. John Wiley & Sons. ISBN 978-1-118-93329-9.

17 "Standard model of particles and interactions". Contemporary Physics Education Project. 2000. Retrieved 2 January 2017.

18 K.A. Olive et al. (Particle Data Group) (2014). "Review of Particle

Physics". Chinese Physics C. 38 (9): 090001.

19 S. Bais (2005), The Equations: Icons of knowledge. p. 84. ISBN 0-674-01967-9.

20 - C. Rovelli (2008), "Quantum gravity". Scholarpedia. 3 (5): 7117.
   - A. Zee (2010), Quantum Field Theory in a Nutshell (second ed.). Princeton University Press. pp. 172,434-435.

21 H. Cavendish, "Experiments to determine the Density of the Earth", Philosophical Transactions of the Royal Society of London, (part II) 88 p.469-526 (21 June 1798), reprinted in Cavendish 1798

22 G. Rosi, F. Sorrentino, L. Cacciapuoti, M. Prevedelli, G.M. Tino, "Precision measurement of the Newtonian gravitational constant using cold atoms ", Nature 510 (2014), 518-521. Schlamminger, Stephan (18 June 2014). "Fundamental constants: A cool way to measure big G". Nature. 510: 478-480.

23 D.R. Williams (16 March 2017), "Earth Fact Sheet". NASA/Goddard Space Flight Center. Retrieved 26 July 2018.

24 - E. Burtt, The Metaphysical Foundations of Modern Physical Science. p. 52.
   - G.J. Holton, S.G. Brush (2001), Physics, the Human Adventure. Rutgers University Press. p. 45. ISBN 0-8135-2908-5.

25 189.R.4.47, ff. 7-8, Trinity College Library, Cambridge, UK

26 E. Verliinde, "On the Origin of Gravity and the Laws of Newton", JHEP. 2011.

27 A. Einstein (November 25, 1915), "Die Feldgleichungen der Gravitation". Sitzungsberichte der Preussischen Akademie der Wissenschaften zu Berlin: 844-847.

28 A. Einstein (1916), "The Foundation of the General Theory of Relativity". Annalen der Physik. 354 (7): 769.

29 F.W. Dyson, A.S. Eddington, C.R. Davidson (1920), "A

Determination of the Deflection of Light by the Sun's Gravitational Field, from Observations Made at the Solar eclipse of May 29, 1919". Phil. Trans. Roy. Soc. A. 220 (571-581): 291-333.

30 K.J. Treschman, "Recent astronomical tests of general relativity", Inter J Phys. Sci. 10(2), pp. 90-105 (2015)

31 K. Schwarzschild, "Über das Gravitationsfeld eines Massenpunktes nach der Einsteinschen Theorie", Sitzungsberichte der Königlich Preussischen Akademie der Wissenschaften 7 (1916) 189-196.

32 E. Verlinde, The Hidden Phase Space of our Universe, Strings 2011, Uppsala, 1 July 2011.

33 L. Randall (2005), Warped Passages: Unraveling the Universe's Hidden Dimensions. Ecco. ISBN 978-0-06-053108-9.

34 - R. P. Feynman, F.B. Morinigo, W.G. Wagner, B. Hatfield (1995), Feynman lectures on gravitation. Addison-Wesley. ISBN 978-0-201-62734-3.
- A. Zee (2003), Quantum Field Theory in a Nutshell. Princeton University Press. ISBN 978-0-691-01019-9.

35 C. Rovelli (August 2008), "Loop Quantum Gravity" (PDF). CERN. Retrieved 14 September 2014. https://indico.cern.ch/event/21917/contributions/1521199/attachments/354819/494246/Rovelli08.pdf

36 The Force of Empty Spaces, Physical Review Focus 3 December 1998.

37 - V. Weisskopf (1935), "Probleme der neueren Quantentheorie des Elektrons". Naturwissenschaften. 23: 631-637.
- T.A. Welton (1948), "Some observable effects of the quantum-mechanical fluctuations of the electromagnetic field". Phys. Rev. 74 (9): 1157.

38 E.T. Jaynes, F.W. Cummings (1963), "Comparison of quantum and semiclassical radiation theories with application to the

beam maser". Proceedings of the IEEE. 51 (1).

39 G. Aruldhas (2009). "§15.15 Lamb Shift". Quantum Mechanics (2nd ed.). Prentice-Hall of India Pvt. Ltd. p. 404. ISBN 81-203-3635-6.

40 S. Carroll, June 22, 2006C-SPAN broadcast of Cosmology at Yearly Kos Science Panel, Part 1.

41 R. Feynman (1985), QED: The Strange Theory of Light and Matter. Princeton University Press.

42 - L. de la Peña, A.M. Cetto, A. Valdes-Hernandez (2014), "The Emerging Quantum: The Physics Behind Quantum Mechanics": 19.
  - L. de la Peña, A.M. Cetto, A. Valdes-Hernandez (2014), "The zero-point field and the emergence of the quantum" . International Journal of Modern Physics E. 23 (09): 1450049.
  - G. Grössing (2014), "Emergence of quantum mechanics from a sub-quantum statistical mechanics". International Journal of Modern Physics B. 28 (26): 1450179
  - A. Valentini (2009), "Beyond the quantum". Physics World. 22 (11): 32-37.
  - G. Musser (November 18, 2013), "Cosmological Data Hint at a Level of Physics Underlying Quantum Mechanics". blogs.scientificamerican.com. Scientific American. Retrieved 5 December 2016.
  - B. Haisch, A. Rueda, H.E. Puthoff (1994), "Inertia as a zero-point-field Lorentz force" . Physical Review A. 49 (2): 678-694.
  - R. Matthews (25 February 1995), "Nothing Like a Vacuum". New Scientist. 145 (1966): 30-33.

43 - P.W. Milonni, "The Quantum Vacuum", An Introduction to QED, Acdemic Press 2013
  - L. de la Peña, A.M. Cetto, "The Quantum Dice: An Introduction to Stochastic Electrodynamics"

44 P. Hoyng, Relativistic Astrophysics and Cosmology: A Primer, Springer-Veralag, Berlin, 274, 2006.

45 https://refractiveindex.info/

46 L.E. Kinsler et al. (2000), Fundamentals of acoustics, 4th Ed., John Wiley and sons Inc., New York, USA.

47 46.

48 J. Krautkrämer, H. Krautkrämer (1990), Ultrasonic testing of materials, 4th fully revised edition, Springer-Verlag, Berlin, Germany, p. 497.

49 B. Moore, T. Jaglinski, D.S. Stone, R.S. Lakes, "Negative incremental bulk modulus in foams", Philosophical Magazine Letters, 86, 651-659, (2006).

50 R. Lakes, K.W. Wojciechowski, "Negative compressibility, negative Poisson's ratio, and stability", Physica Status Solidi, 245, No. 3, 545-551, Feb 4 (2008).

51 S. Naifeh, G.W. Smith (2011), Van Gogh: The Life. New York: Random House. ISBN 978-0-375-50748-9.

52 R.D. Reasenberg, B.R. Patla, J.D. Phillips, R. Thapa, "Design and characteristics of a WEP test in a sounding-rocket payload", Classical and Quantum Gravity 27, 095005 (2010).

53 W. Rindler (2006), Relativity: Special, General, And Cosmological. Oxford University Press. p. 22.

54 M. Jammer, Concepts of Mass in Classical and Modern Physics, (Dover, 1997), p 203.

55 R.V. Eötvös, D. Pekár, E. Fekete (1922), "Beiträge zum Gesetz der Proportionalität von Trägheit und Gravität". Annalen der Physik. 68: 11-66

56 J. Overduin, F. Everitt, J. Mester, P. Worden (2009), "The Science Case for STEP". Advances in Space Research. 43 (10): 1532-1537.

57 H. Nishino et al. (2009), "Search for Proton Decay via p→e+π⁰ and p→μ+π⁰ in a Large Water Cherenkov Detector". Physical Review Letters. 102 (14): 141801.

58 R. Feynman, QED: The Strange Theory of Light and Matter, Penguin 1990 Edition, p 84.

59 C.J. Davisson, L.H. Germer (1 April 1928), "Reflection of Electrons by a Crystal of Nickel". Proceedings of the National Academy of Sciences of the United States of America. 14 (4): 317-322.

60 M. Arndt, O. Nairz, J. Voss-Andreae, C. Keller, G. van der Zouw, A. Zeilinger (14 October 1999). "Wave-particle duality of C60". Nature. 401 (6754): 680-682.

61 J. Manners (2000), Quantum Physics: An Introduction, CRC Press, pp. 53-56, ISBN 978-0-7503-0720-8.

62 A. Beiser, Concepts of Modern Physics, 6th edition, McGraw-Hill 2003, p 97.

63 - L. Randall, Warped Passages: Unraveling the Mysteries of the Universe's Hidden Dimensions, p.286.
- P.A. Tipler, R.A. Llewellyn (2008), Modern Physics (5th ed.). New York: W.H. Freeman & Co. p. 54. ISBN 978-0-7167-7550-8.

64 - R. Jia, G. Amulele, P.V. Zinin, S. Odake, P. Eng, V. Khabashesku, W.L. Maoe, C.M. Li, Elastic and inelastic behavior of graphitic C_3N_4 under high pressure", Chemical Physics Letters 575 (2013) 67-70.
- Y. Zou, X. Wang, T. Chen, X. Li, X. Qi, D. Welch, P. Zhu, B. Liu, T. Cui, B. Li, "Hexagonal-structured ε-NbN: ultra-incompressibility, high shear rigidity, and a possible hard superconducting material", Scientific Reports June 2015.

65 Wertz, edited by J.R. Wertz and W.J. Larson (2010), Space mission analysis and design (3rd ed.). Hawthorne, Calif.: Microcosm. p. 135. ISBN 978-1881883-10-4.

66 A.S. Eddington (1919), "The total eclipse of 1919 May 29 and

the influence of gravitation on light". Observatory 42:119-122.

67 W.W. Campbell, R. Trumpler (1928), "Observations made with a pair of five-foot cameras on the light-deflections in the Sun's gravitational field at the total solar eclipse of September 21, 1922". Lick Observatory Bulletin 397(13):130-160.

68 K.J. Treschman, "Recent astronomical tests of general relativity", International Journal of Physical Sciences, Vol. 10(2), pp. 90-105, 30 Janaury, 2015.

69 U. Le Verrier (1859), (in French), "Lettre de M. Le Verrier á M. Faye sur la théorie de Mercure et sur le mouvement du périhélie de cette planète", Comptes rendus hebdomadaires des séances de l'Acadèmie des sciences (Paris), vol. 49 (1859), pp.379-383.

70 G.M. Clemence (1947), "The Relativity Effect in Planetary Motions". Reviews of Modern Physics. 19 (4): 361-364. Bibcode:1947RvMP...19..361C. doi:10.1103/RevModPhys.19.361.

71 R.S. Park et al. "Precession of Mercury's Perihelion from Ranging to the MESSENGER Spacecraft." The Astronomical Journal 153.3 (2017): 121.

72 A. Biswas, K.R.S. Mani (2008), "Relativistic perihelion precession of orbits of Venus and the Earth". Central European Journal of Physics. vl. 6 (3): 754-758.

73 J.M. Weisberg, J.H. Taylor (July 2005), "The Relativistic Binary Pulsar B1913+16: Thirty Years of Observations and Analysis". Written at San Francisco. In F.A. Rasio; I.H. Stairs. Binary Radio Pulsars. ASP Conference Series. 328. Aspen, Colorado, USA: Astronomical Society of the Pacific. p. 25

74 S. Hawking, On the Shoulders of Giants : the Great Works of Physics and Astronomy. Philadelphia, Pennsylvania, USA: Running Press. pp. 1243, Foundation of the General Relativity (translated from Albert Einstein's Die Grundlage der Allgemeine Relativitätstheorie, first published in 1916 in Annalen der

Physik, volume 49). ISBN 0-7624-1348-4.

75 - Y. Friedman, S. Livshitz, J. M. Steiner, "Predicting the relativistic periastron advance of a binary without curving spacetime", Eur. Phys. Lett., 116 (2016) 19001
- Y. Friedman and J. M. Steiner, "Predicting Mercury's Precession using Simple Relativistic Newtonian Dynamics", Eur. Phys. Lett. (EPL), 113 (2016) 39001

76 B. Cordero, V. Gómez, A.E. Platero-Prats, M. Revés, J. Echeverría, E. Cremades, F. Barragán, S. Alvarez (2008), "Covalent radii revisited". Dalton Trans. (21): 2832-2838.

77 K. Lodders (10 July 2003), "Solar System Abundances and Condensation Temperatures of the Elements" (PDF). The Astrophysical Journal. 591 (2): 1220-1247.

78 R. García et al. (2007), "Tracking solar gravity modes: the dynamics of the solar core". Science. 316 (5831): 1591-1593.

79 "NASA/Marshall Solar Physics". Marshall Space Flight Center. 18 January 2007. Retrieved 11 July 2009.

80 R. García (2007), "Tracking solar gravity modes: the dynamics of the solar core". 《Science》 316 (5831): 1591-1593.

81 World Book at NASA. NASA. Archived from the original on 10 May 2013. Retrieved 10 October 2012.

82 C. Broggini (2003), Physics in Collision, Proceedings of the XXIII International Conference: Nuclear Processes at Solar Energy. XXIII Physics in Collisions Conference. Zeuthen, Germany. p. 21.

83 K. Nakamura (2010), "Review of Particle Physics". Journal of Physics G: Nuclear and Particle Physics. 37 (7A): 075021.

84 "Solar System Exploration: Planets: Sun: Facts & Figures". NASA. Archived from the original on 2 January 2008.

85

http://earthsky.org/space/suns-core-rotates-4x-faster-than-surf ace, "Sun's core rotates 4x faster than surface" By Deborah Byrd in Space | August 2, 2017

86
http://starchild.gsfc.nasa.gov/docs/StarChild/questions/question 18.html NASA - StarChild Question of the Month for February 2000

87 C. Francis, E. Anderson (June 2014), "Two estimates of the distance to the Galactic Centre". Monthly Notices of the Royal Astronomical Society. 441 (2): 1105-1114.

88 O.Y. Gnedin (2010), "The mass profile of the Galaxy to 80 kpc". The Astrophysical Journal. 720: L108-L112.

89 J. Kormendy, L. Ho (2013), "Coevolution (Or Not) of Supermassive Black Holes and Host Galaxies". Annual Review of Astronomy and Astrophysics,. 51: 511-653.

90 D.R. Williams (1 September 2004), "Earth Fact Sheet". NASA. Retrieved 9 August 2010.

91 T.H. Jordan (1979), "Structural geology of the Earth's interior". Proceedings of the National Academy of Sciences of the United States of America. 76 (9): 4192-4200.

92 T. Tanimoto (1995), "Crustal Structure of the Earth" . In Thomas J. Ahrens. Global Earth Physics: A Handbook of Physical Constants. Washington, DC: American Geophysical Union. ISBN 0-87590-851-9. Archived from the original on 16 October 2006. Retrieved 3 February 2007.

93 R.A. Kerr (2005), "Earth's Inner Core Is Running a Tad Faster Than the Rest of the Planet". Science. 309 (5739): 1313a.

94 International Earth Rotation and Reference Systems Service (IERS) Working Group (2004). "General Definitions and Numerical Standards". In D.D. McCarthy, G. Petit, IERS Conventions (2003). IERS Technical Note No. 32. Frankfurt am Main: Verlag des Bundesamts für Kartographie und Geodäsie. p. 12. ISBN

3-89888-884-3. Retrieved 29 April 2016.

95 P.K. Seidelmann, B.A. Archinal, M.F. A'Hearn et al. (2007), "Report of the IAU/IAG Working Group on cartographic coordinates and rotational elements: 2006". Celestial Mechanics and Dynamical Astronomy. 98 (3): 155-180.

96 A.N. Cox, ed. (2000). Allen's Astrophysical Quantities (4th ed.). New York: AIP Press. p. 244. ISBN 978-0-387-98746-0. Retrieved 17 August 2010.

97 - R. Boynton (2001), "Precise Measurement of Mass". Sawe Paper No. 3147. Arlington, Texas: S.A.W.E., Inc. Retrieved 2007-01-21.
    - "Curious About Astronomy?", Cornell University, retrieved June 2007

98 http://www.hani.co.kr/arti/science/science_general/176024.html, 미래&과학 2006.12.04. 21억년 뒤 하루는 30시간!? 유후~

99 - D. Gautier, B. Conrath, M. Flasar, R. Hanel, V. Kunde, A. Chedin, N. Scott (1981), "The helium abundance of Jupiter from Voyager". Journal of Geophysical Research. 86 (A10): 8713-8720.
    2. Kunde, V. G.; et al. (September 10, 2004). "Jupiter's Atmospheric Composition from the Cassini Thermal Infrared Spectroscopy Experiment". Science. 305 (5690): 1582-86.

100 T. Guillot, D.J. Stevenson, W.B. Hubbard, D. Saumon (2004), "Chapter 3: The Interior of Jupiter". The Planet, Satellites and Magnetosphere. Cambridge University Press. ISBN 0-521-81808-7.

101 T. Guillot, D.J. Stevenson, W.B. Hubbard (1997), "New Constraints on the Composition of Jupiter from Galileo Measurements and Interior Models". Icarus. 130 (2): 534-539.

102 101.

103 100.

104 L.-A McFadden, P. Weissman, T. Johnson, eds. Encyclopedia

of the Solar System (2nd ed.). Academic Press. p. 412. ISBN 0-12-088589-1.

105 G. J. I. Patrick (2003), Giant Planets of Our Solar System: Atmospheres, Composition, and Structure. Springer. ISBN 3-540-00681-8.

106 "Kepler's Supernova: Recently Observed Supernova". Universe for Facts. Retrieved 21 December 2014.

107 P.F. Winkler, G. Gupta, K.S. Long (2003), "The SN 1006 Remnant: Optical Proper Motions, Deep Imaging, Distance, and Brightness at Maximum". Astrophysical Journal. 585 (1): 324.

108 S. Dong et al. (2016). "ASASSN-15lh: A highly super-luminous supernova". Science. 351 (6270): 257-260.

109 A. Heger, C.L. Fryer, S.E. Woosley, N. Langer, D.H. Hartmann (2003), "How Massive Single Stars End Their Life". Astrophysical Journal. 591: 288.

110 K. Schawinski et al. (2008). "Supernova Shock Breakout from a Red Supergiant". Science. 321 (5886): 223-226.

111 F.X. Timmes, S.E. Woosley, T.A. Weaver (1995), "Galactic chemical evolution: Hydrogen through zinc". Astrophysical Journal Supplement Series. 98: 617.

112 F.B. Bianco, M. Modjaz, M. Hicken, A. Friedman, R.P. Kirshner, J.S. Bloom, P. Challis, G.H. Marion, W.M. Wood-Vasey, A. Rest (2014), "Multi-color Optical and Near-infrared Light Curves of 64 Stripped-envelope Core-Collapse Supernovae". The Astrophysical Journal Supplement. 213 (2): 19.

113 A.L. Piro, T.A. Thompson, C.S. Kochanek (2014), "Reconciling 56Ni production in Type Ia supernovae with double degenerate scenarios". Monthly Notices of the Royal Astronomical Society. 438 (4): 3456.

114 F.K. Röpke, W. Hillebrandt (2004), "The case against the progenitor's carbon-to-oxygen ratio as a source of peak

luminosity variations in Type Ia supernovae". Astronomy and Astrophysics Letters. 420 (1): L1-L4.

115 B. Paczyński (1976), "Common Envelope Binaries". In Eggleton, P.; Mitton, S.; Whelan, J. Structure and Evolution of Close Binary Systems. IAU Symposium No. 73. Dordrecht: D. Reidel. pp. 75-80.

116 W. Hillebrandt, J.C. Niemeyer (2000), "Type IA Supernova Explosion Models". Annual Review of Astronomy and Astrophysics. 38 (1): 191-230.

117 - A. Heger, C.L. Fryer, S.E. Woosley, N. Langer, D.H. Hartmann (2003), "How Massive Single Stars End Their Life". Astrophysical Journal. 591: 288.
    - K. Nomoto, M. Tanaka, N. Tominaga, K. Maeda (2010), "Hypernovae, gamma-ray bursts, and first stars". New Astronomy Reviews. 54 (3-6): 191.

118 A.J.T. Poelarends, F. Herwig, N. Langer, A. Heger (2008), "The Supernova Channel of Super-AGB Stars". The Astrophysical Journal. 675: 614.

119 - 117 (Heger et al.).
    - G. Faure, T.M. Mensing (2007), "Life and Death of Stars". Introduction to Planetary Science. pp. 35-48.

120 G. Gilmore (2004), "The Short Spectacular Life of a Superstar". Science. 304 (5697): 1915-1916.

121 D. Bodansky, D.D. Clayton, W.A. Fowler (1968), "Nucleosynthesis During Silicon Burning". Physical Review Letters. 20 (4): 161.

122 D. Kasen, S.E. Woosley (2009), "Type Ii Supernovae: Model Light Curves and Standard Candle Relationships". The Astrophysical Journal. 703 (2): 2205.

123 S.M. Matz, G.H. Share, M.D. Leising, E.L. Chupp, W.T. Vestrand, W.R. Purcell, M.S. Strickman, C. Reppin (1988), "Gamma-ray line emission from SN 1987A". Nature. 331 (6155):

416.

124 P.A. Mazzali, K.I. Nomoto, E. Cappellaro, T. Nakamura, H. Umeda, K. Iwamoto (2001), "Can Differences in the Nickel Abundance in Chandrasekhar-Mass Models Explain the Relation between the Brightness and Decline Rate of Normal Type Ia Supernovae?". The Astrophysical Journal. 547 (2): 988.

125 H. Mes, I. Ahmad, J. Hebert (1966), "Progress of theoretical physics: Resonance in the Nucleus". Institute of Physics. Ottawa, Canada: University of Ottawa (Department of Physics). 3 (3): 556-567.

126 Y.-Z. Qian, P. Vogel, G.J. Wasserburg (1998), "Diverse Supernova Sources for the r-Process". Astrophysical Journal. 494 (1): 285-296.

127 D.P. Cox (1972), "Cooling and Evolution of a Supernova Remnant". Astrophysical Journal. 178: 159.

128 A.G.W. Cameron, J.W. Truran (1977), "The supernova trigger for formation of the solar system". Icarus. 30 (3): 447.

129 A. Melott et al. (2004). "Did a gamma-ray burst initiate the late Ordovician mass extinction?". International Journal of Astrobiology. 3 (2): 55-61.

130 - B.D. Fields, K.A. Hochmuth, J. Ellis (2005), "Deep-Ocean Crusts as Telescopes: Using Live Radioisotopes to Probe Supernova Nucleosynthesis". The Astrophysical Journal. 621 (2): 902.
    - K. Knie et al. (2004), "^{60}Fe Anomaly in a Deep-Sea Manganese Crust and Implications for a Nearby Supernova Source". Physical Review Letters. 93 (17): 171103-171106.
    - B.D. Fields, J. Ellis (1999), "On Deep-Ocean Fe-60 as a Fossil of a Near-Earth Supernova". New Astronomy. 4 (6): 419-430.

131 D. Wonnacott, B.J. Kellett, D.J. Stickland (1993), "IK Peg - A nearby, short-period, Sirius-like system". Monthly Notices of

the Royal Astronomical Society. 262 (2): 277-284

132 J. Bally, B. Reipurth (2006), The Birth of Stars and Planets (illustrated ed.). Cambridge University Press. p. 207

133 M.C. Miller, "Introduction to neutron stars". http://www.astro.umd.edu/~miller/nstar.html.

134 A. Reisenegger, "Origin and Evolution of Neutron Star Magnetic Fields". Universidade Federal do Rio Grande do Sul. Retrieved 21 March 2016.

135 J. Hessels, S.M. Ransom, I.H. Stairs, P.C.C. Freire et al. (2006), "A Radio Pulsar Spinning at 716 Hz". Science. 311 (5769): 1901-1904.

136 W. Baade, F. Zwicky (1934), "Remarks on Super-Novae and Cosmic Rays". Physical Review. 46 (1): 76-77.

137 J. Chadwick (1932), "On the possible existence of a neutron". Nature. 129 (3252): 312

138 K. Lang (2007), A Companion to Astronomy and Astrophysics: Chronology and Glossary with Data Tables (illustrated ed.). Springer Science & Business Media.

139 J.M. Weisberg, D.J. Nice, J.H. Taylor (20 October 2010), "Timing Measurements of the Relativistic Binary Pulsar PSR B1913+16". The Astrophysical Journal. 722 (2): 1030-1034.

140 "Multi-messenger Observations of a Binary Neutron Star Merger". The Astrophysical Journal Letters. 848 (2): L12. 2017. Retrieved 16 October 2017.

141 - P.B. Demorest, T. Pennucci, S.M. Ransom, M.S. Roberts et al. (2010), "A two-solar-mass neutron star measured using Shapiro delay". Nature. 467 (7319): 1081-1083.
    2. Antoniadis, J (2012). "A Massive Pulsar in a Compact Relativistic Binary". Science. 340 (6131): 1233232.

142 R.M. Wald, "Black Holes" in General Relativity, University of

Chicago Press, Chicago and London, 1984.

143 R.M. Wald (1997), "Gravitational Collapse and Cosmic Censorship"

144 - B.P. Abbott et al. (2016). "Observation of Gravitational Waves from a Binary Black Hole Merger". Phys. Rev. Lett. 116 (6): 061102.
   - D. Overbye (15 June 2016), "Scientists Hear a Second Chirp From Colliding Black Holes". New York Times. Archived from the original on 15 June 2016. Retrieved 15 June 2016.

145 - B.F. Schutz (2003), Gravity from the ground up. Cambridge University Press. p. 110.
   - P.C.W. Davies (1978), "Thermodynamics of Black Holes" . Reports on Progress in Physics. 41 (8): 1313-1355.

146 P. Srikanta (2017-03-09), Recent Developments in Intelligent Nature-Inspired Computing. IGI Global. ISBN 9781522523239.

147 "NASA's NuSTAR Sees Rare Blurring of Black Hole Light". NASA. 12 August 2014. Archived from the original on 13 August 2014. Retrieved 12 August 2014.

148 J.E. McClintock, R.A. Remillard (2006), "Black Hole Binaries". In W. Lewin, M. van der Klis, Compact Stellar X-ray Sources. Cambridge University Press.

149 J.A. Marck, "Short-cut method of solution of geodesic equations for Schwarzchild black hole", Class. Quant. Grav. 13 (1996) 393-402.

150 A. Celotti, J.C. Miller, D.W. Sciama (1999), "Astrophysical evidence for the existence of black holes". Classical and Quantum Gravity. 16 (12A): A3-A21.

151 - "Hubble directly observes the disc around a black hole". www.spacetelescope.org. Archived from the original on 8 March 2016. Retrieved 7 March 2016.
   - J.A. Muñoz, E. Mediavilla, C.S. Kochanek, E. Falco, A.M. Mosquera (2011-12-01), "A Study of Gravitational Lens

Chromaticity with the Hubble Space Telescope". The Astrophysical Journal. 742 (2): 67.

152 S.W. Hawking (1971), "Gravitational Radiation from Colliding Black Holes". Physical Review Letters. 26 (21): 1344-1346.

153 R.M. Wald (2001), "The Thermodynamics of Black Holes". Living Reviews in Relativity. 4: 6.

154 G. 't Hooft (2001), "The Holographic Principle". In Zichichi, A. Basics and highlights in fundamental physics. Subnuclear series. 37. World Scientific.

155 S.D. Mathur (2011), The information paradox: conflicts and resolutions. XXV International Symposium on Lepton Photon Interactions at High Energies.

156 Z. Merali (4 April 2013), "Astrophysics: Fire in the hole!". Nature. pp. 20-23.

157 A. Celotti, J.C. Miller, D.W. Sciama (1999), "Astrophysical evidence for the existence of black holes". Class. Quant. Grav. 16 (12A): A3-A21.

158 P. Srikanta (2017-03-09), Recent Developments in Intelligent Nature-Inspired Computing. IGI Global. ISBN 9781522523239.

159 K. Schwarzschild, Über das Gravitationsfeld eines Massenpunktes nach der Einsteinschen Theorie, Sitzungsberichte der Königlich Preussischen Akademie der Wissenschaften 7 (1916) 189-196.

160 J. Earman, "The Penrose-Hawking singularity theorems: History and Implications", In: The expanding worlds of general relativity, H. Goenner, J. Renn, J. Ritter, T. Sauer (Eds.). Birk presentations of the fourth conference on the and gravitation held in Berlin, 1999, p. 235-267. ISBN 978-0-8176-4060-6

161 E.L. Wright (2004), "Theoretical Overview of Cosmic Microwave Background Anisotropy". In W. L. Freedman. Measuring and Modeling the Universe. Carnegie Observatories Astrophysics

Series. Cambridge University Press. p. 291.

162 A.H. Guth (1998), The Inflationary Universe: The Quest for a New Theory of Cosmic Origins. Basic Books. p. 186. ISBN 978-0201328400. OCLC 35701222.

163 D. Cirigliano, H.J. de Vega, N.G. Sanchez (2005), "Clarifying inflation models: The precise inflationary potential from effective field theory and the WMAP data". Physical Review D. 71 (10): 77-115.

164 B. Abbott (2007), "Microwave (WMAP) All-Sky Survey". Hayden Planetarium. Archived from the original on 2013-02-13. Retrieved 2008-01-13.

165 E. Gawiser, J. Silk (2000), "The cosmic microwave background radiation". Physics Reports. 333-334: 245-267. ˘

166 M. Kaku (2014), "First Second of the Big Bang". How the Universe Works. Discovery Science.

167 G.F. Smoot (2006), "Cosmic Microwave Background Radiation Anisotropies: Their Discovery and Utilization". Nobel Lecture. Nobel Foundation. Retrieved 2008-12-22.

168 A. Unsöld, B. Bodo (2002), The New Cosmos, An Introduction to Astronomy and Astrophysics (5th ed.). Springer-Verlag. p. 485. ISBN 3-540-67877-8.

169 R.A. Alpher, R.C. Herman (1948), "On the Relative Abundance of the Elements". Physical Review. 74 (12): 1737-1742.

170 A.G. Doroshkevich, I.D. Novikov (1964), "Mean Density of Radiation in the Metagalaxy and Certain Problems in Relativistic Cosmology". Soviet Physics Doklady. 9 (23): 4292-4298.

171 The Cosmic Microwave Background Radiation (Nobel Lecture) by R. Wilson 8 Dec 1978, p. 474.

172 L. Pogosian et al. (2003). "Observational constraints on cosmic string production during brane inflation". Physical Review D. 68

(2): 023506.

173 C.L. Bennett, (WMAP collaboration) G. Hinshaw, et al. (2003), "First-year Wilkinson Microwave Anisotropy Probe (WMAP) observations: preliminary maps and basic results". Astrophysical Journal Supplement Series. 148: 1-27.

174 - W. Clavin, J.D. Harrington (21 March 2013), "Planck Mission Brings Universe Into Sharp Focus". NASA. Retrieved 21 March 2013.
    - Staff (21 March 2013), "Mapping the Early Universe". New York Times. Retrieved 23 March 2013.

175 - G. Rossmanith, C. Räth, A.J. Banday, G. Morfill (2009), "Non-Gaussian Signatures in the five-year WMAP data as identified with isotropic scaling indices". Monthly Notices of the Royal Astronomical Society. 399 (4): 1921-1933.
    - A. Bernui, B. Mota, M.J. Rebouças, R. Tavakol (2005), "Mapping the large-scale anisotropy in the WMAP data". Astronomy and Astrophysics. 464 (2): 479-485.
    - T.R. Jaffe, A.J. Banday, H.K. Eriksen, K.M. Górski, F.K. Hansen (2005), "Evidence of vorticity and shear at large angular scales in the WMAP data: a violation of cosmological isotropy?". The Astrophysical Journal. 629: L1-L4.

176
    https://www.newscientist.com/article/dn23301-planck-shows-alm ost-perfect-cosmos-plus-axis-of-evil/

177 - H. Liu, T.-P. Li (2009), "Improved CMB Map from WMAP Data".
    - U. Sawangwit, T. Shanks (2010), "Lambda-CDM and the WMAP Power Spectrum Beam Profile Sensitivity".
    - H. Liu et al. (2010). "Diagnosing Timing Error in WMAP Data".

178
    http://background.uchicago.edu/~whu/intermediate/baryons.htm l, Wayne Hu. "Baryons and Inertia".

179
http://background.uchicago.edu/~whu/intermediate/driving.html,
Wayne Hu. "Radiation Driving Force"

180 W. Hu, M. White (1996), "Acoustic Signatures in the Cosmic
Microwave Background". Astrophysical Journal. 471: 30-51.

181 WMAP Collaboration; L. Verde, H.V. Peiris, E. Komatsu, M.R.
Nolta, C.L. Bennett, M. Halpern, G. Hinshaw et al. (2003),
"First-Year Wilkinson Microwave Anisotropy Probe (WMAP)
Observations: Determination of Cosmological Parameters".
Astrophysical Journal Supplement Series. 148 (1): 175-194.

182 G. Pagliaroli, F. Vissani, M.L. Costantini, A. Ianni (2009),
"Improved analysis of SN 1987A antineutrino events".
Astroparticle Physics. 31 (3): 163-176.

183 Cosmic Neutrinos Detected, Confirming The Big Bang's Last
Great Prediction - Forbes coverage of original paper: First
Detection of the Acoustic Oscillation Phase Shift Expected from
the Cosmic Neutrino Background - Follin, Knox, Millea, Pan,
pub. Phys. Rev. Lett. 26 August 2015.

184 "Dark Energy, Dark Matter". NASA Science: Astrophysics. 5
June 2015.

185 M. Persic, P. Salucci (1 September 1992), "The baryon content
of the Universe". Monthly Notices of the Royal Astronomical
Society. 258 (1): 14P-18P.

186 V. Trimble (1987), "Existence and nature of dark matter in the
universe". Annual Review of Astronomy and Astrophysics. 25:
425-472.

187 C.J. Copi, D.N. Schramm, M.S. Turner (1995), "Big-Bang
Nucleosynthesis and the Baryon Density of the Universe".
Science. 267 (5195): 192-199.

188 - F. Zwicky (1933), "Die Rotverschiebung von extragalaktischen
Nebeln", Helvetica Physica Acta, 6: 110-127.
   - F. Zwicky (1937), "On the Masses of Nebulae and of Clusters

of Nebulae", Astrophysical Journal, 86: 217.

189 E. Corbelli, P. Salucci (2000), "The extended rotation curve and the dark matter halo of M33". Monthly Notices of the Royal Astronomical Society. 311 (2): 441-447.

190 F. Zwicky (February 1937), "Nebulae as Gravitational Lenses", Physical Review, 51 (4): 290,

191 D. Walsh, R.F. Carswell, R.J. Weymann (May 31, 1979), "0957 + 561 A, B - Twin quasistellar objects or gravitational lens", Nature, 279 (5712): 381-384

192 A.N. Taylor et al. (1998), "Gravitational Lens Magnification and the Mass of Abell 1689". The Astrophysical Journal. 501 (2): 539-553.

193 X. Wu, T. Chiueh, L. Fang, Y. Xue (1998), "A comparison of different cluster mass estimates: consistency or discrepancy?". Monthly Notices of the Royal Astronomical Society. 301 (3): 861-871.

194 A. Refregier (2003), "Weak gravitational lensing by large-scale structure". Annual Review of Astronomy and Astrophysics. 41 (1): 645-668.

195 W. Hu. "Baryons and Inertia".

http://background.uchicago.edu/~whu/intermediate/baryons.html

196 W. Hu. "Radiation Driving Force".
http://background.uchicago.edu/~whu/intermediate/driving.html

197 D. Merritt (February 2017), "Cosmology and convention". Studies in History and Philosophy of Science Part B. 57 (2017): 41-52.

198 P.J.E. Peebles, B. Ratra (2003), "The cosmological constant and dark energy". Reviews of Modern Physics. 75 (2): 559-606.

199 - P.J. Steinhardt, N. Turok (2006), "Why the cosmological constant is small and positive". Science. 312 (5777): 1180-1183.

200 R.R. Caldwell, R. Dave, P.J. Steinhardt (1998), "Cosmological Imprint of an Energy Component with General Equation-of-State". Phys. Rev. Lett. 80 (8): 1582-1585.

201 H. Kragh (2012), "Preludes to dark energy: zero-point energy and vacuum speculations". Archive for History of Exact Sciences. Volume 66, Issue 3, pp 199-240

202 A.S. Eddington (5.1930), "On the instability of Einstein's spherical world". 《Monthly Notices of the Royal Astronomical Society》 90: 668-678.

203 - A.G. Riess et al. (1998). "Observational evidence from supernovae for an accelerating universe and a cosmological constant". Astronomical Journal. 116 (3): 1009-38.
    - S. Perlmutter et al. (1999). "Measurements of Omega and Lambda from 42 high redshift supernovae". Astrophysical Journal. 517 (2): 565-86.

204 Z.-Y. Wang (2016), "Modern Theory for Electromagnetic Metamaterials". Plasmonics. 11 (2): 503-508.

205 "Universe 101". NASA. Retrieved September 9, 2015. "The actual density of atoms is equivalent to roughly 1 proton per 4 cubic meters."

206 D.N. Spergel (WMAP collaboration) et al. (March 2006), "Wilkinson Microwave Anisotropy Probe (WMAP) three year results: implications for cosmology".

207 M. Kowalski et al. (October 27, 2008), "Improved Cosmological Constraints from New, Old and Combined Supernova Datasets". The Astrophysical Journal. Chicago: University of Chicago Press. 686 (2): 749-778.

208 "Big Bang's afterglow shows universe is 80 million years older than scientists first thought". The Washington Post. Archived from the original on 22 March 2013. Retrieved 22 March 2013.

209 S. Carroll (2001), "The cosmological constant". Living Reviews in Relativity. 4.

210 J. Wess, J. Bagger, Supersymmetry and Supergravity. Series: Princeton Series in Physics, Princeton University Press; Revised edition (March 3, 1992) ISBN-10: 0691025304, ISBN-13: 978-0691025308

211 - D.L. Wiltshire (2007), "Exact Solution to the Averaging Problem in Cosmology". Physical Review Letters. 99 (25): 251101.
    - M. Ishak, J. Richardson, D. Garred, D. Whittington, A. Nwankwo, R. Sussman (2007), "Dark Energy or Apparent Acceleration Due to a Relativistic Cosmological Model More Complex than FLRW?". Physical Review D. 78 (12): 123531.
    - T. Mattsson (2007), "Dark energy as a mirage". Gen. Rel. Grav. 42 (3): 567-599.
    - T. Clifton, P. Ferreira (April 2009), "Does Dark Energy Really Exist?". Scientific American. 300 (4): 48-55.

212 - D. Wiltshire (2008), "Cosmological equivalence principle and the weak-field limit". Physical Review D. 78 (8): 084032.
    - S. Gray, "Dark questions remain over dark energy". ABC Science Australia. Retrieved 27 January 2013.
    - Z. Merali (March 2012), "Is Einstein's Greatest Work All Wrong-Because He Didn't Go Far Enough?". Discover magazine. Retrieved 27 January 2013.

213 - J.T. Nielsen, A. Guffanti, S. Sarkar (21 October 2016), "Marginal evidence for cosmic acceleration from Type Ia supernovae". Nature Scientific Reports. 6: 35596.
    - S. Gillespie (21 October 2016), "The universe is expanding at an accelerating rate - or is it?". University of Oxford - News & Events - Science Blog (WP:NEWSBLOG).

214 - J. Silk (2009), Horizons of Cosmology. Templeton Press. p. 208.
    - S. Singh (2005), Big Bang: The Origin of the Universe. Harper Perennial. p. 560.
    - E.J. Wollack (10 December 2010), "Cosmology: The Study of

the Universe". Universe 101: Big Bang Theory. NASA.

215 A. Liddle, An Introduction to Modern Cosmology (2nd ed.). London: Wiley, 2003.

216 - M. Lachieze-Rey, J.-P. Luminet (1995), "Cosmic Topology", Physics Reports, 254 (3): 135-214.
   - G.F.R. Ellis. H. van Elst (1999), "Cosmological models (Cargèse lectures 1998)". In Marc Lachize-Rey. Theoretical and Observational Cosmology. NATO Science Series C. 541. pp. 1-116.

217 A.H. Jaffe. "Cosmology 2012: Lecture Notes".

218 L.F. Low (12 October 2016), ""Constraints on the composite photon theory". Modern Physics Letters A.

219 C.M. Carlisle, Planck Upholds Standard Cosmology, Sky & Telescope, February 10, 2015
http://www.skyandtelescope.com/astronomy-news/planck-upholds-standard-cosmology-0210201523/

220 A. Coc, E. Vangioni (2017), "Primordial nucleosynthesis". International Journal of Modern Physics E. 26 (8): 1741002.

221 Susanna Kohler (15 February 2017), Fixing the Big Bang Theory's Lithium Problem, http://aasnova.org/2017/02/15/fixing-the-big-bang-theorys-lithium-problem/

222 T. Kawabata et al., "Time-Reversal Measurement of the p-Wave Cross Sections of the 7Be(n,α)4He Reaction for the Cosmological Li Problem", Phys. Rev. Lett. 118, 052701 - Published 3 February 2017.

223 M. Rini (2017), "Synopsis: Tackling the Small-Scale Crisis". Physical Review D. 95 (12).

224 D. Merritt, "Cosmology and Convention", Studies In History and Philosophy of Science Part B: Studies In History and Philosophy of Modern Physics, 57(1):41-52, February 2017.

www.ingramcontent.com/pod-product-compliance
Lightning Source LLC
Chambersburg PA
CBHW070528220526
45467CB00003B/900

* 9 7 8 1 7 1 3 0 4 2 0 2 0 *